The Evolving Animal Orchestra

The Evolving Animal Orchestra

In Search of What Makes Us Musical

Henkjan Honing

translated by Sherry Macdonald

The MIT Press
Cambridge, Massachusetts
London, England

First published in Dutch as *Aap slaat maat. Op zoek naar de oorsprong van muzikaliteit bij mens en dier* (Amsterdam, the Netherlands: Nieuw Amsterdam, 2018).

The writing of this book was made possible in part by a Distinguished Lorentz Fellowship from the (Dutch) Lorentz Center for the Sciences and the Netherlands Institute for Advanced Study in the Humanities and Social Sciences (NIAS), with additional support from the Netherlands Organisation for Scientific Research (NWO).

This publication was made possible with financial support from the Dutch Foundation for Literature.

Nederlands letterenfonds
dutch foundation
for literature

This book was set in Stone Serif by Westchester Publishing Services. Printed and bound in the United States of America.

Library of Congress Cataloging-in-Publication Data
Names: Honing, Henkjan, author.
Title: The evolving animal orchestra : in search of what makes us musical / Henkjan Honing ; translated by Sherry Macdonald.
Other titles: Op zoek naar wat ons muzikale dieren maakt. English
Description: Cambridge, MA : The MIT Press, [2019] | Includes bibliographical references and index.
Identifiers: LCCN 2018019051 | ISBN 9780262039321 (hardcover : alk. paper)
Subjects: LCSH: Musical ability. | Music—Physiological aspects. | Musical perception. | Music—Origin.
Classification: LCC ML3820 .H5613 2019 | DDC 781.1/1—dc23 LC record available at https://lccn.loc.gov/2018019051

10 9 8 7 6 5 4 3 2 1

For Anne-Marie

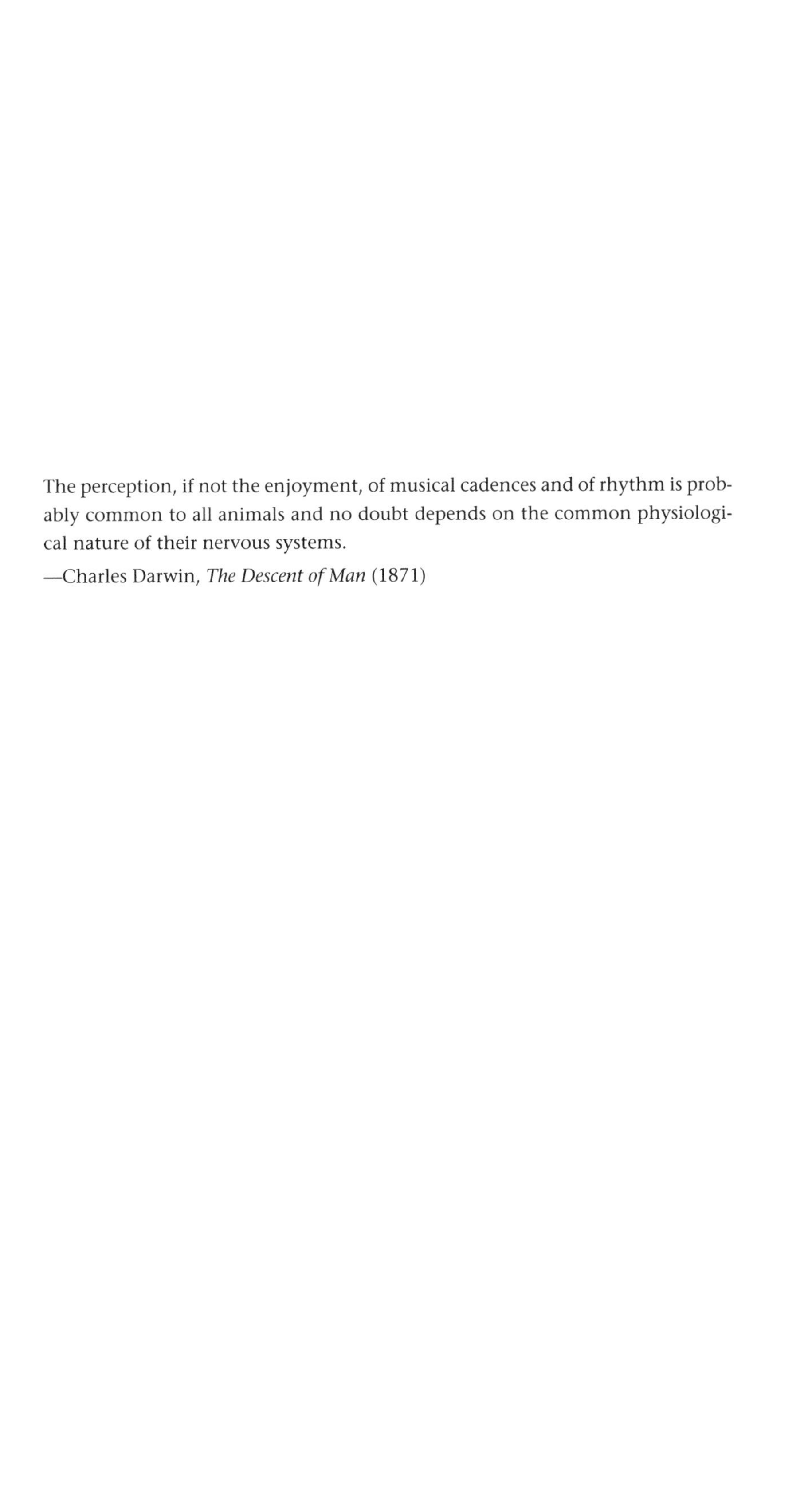

The perception, if not the enjoyment, of musical cadences and of rhythm is probably common to all animals and no doubt depends on the common physiological nature of their nervous systems.

—Charles Darwin, *The Descent of Man* (1871)

Contents

Preface

The term "musicality" probably makes most people think first of their favorite pop star, a preferred composer, or child prodigies like Lang Lang or the Dutch Jussen brothers. The more virtuoso, the better. But musicality is much more than that. It is the musical ability that nearly all human beings possess: a set of traits that allows us to enjoy music. Is musicality so special, though, if we are all musical? Or is it mostly special because we appear to be the only animals with such a vast musical repertoire? Is our musical predisposition unique, like our linguistic ability? Or is musicality something with a long evolutionary history that we share with other animals?

Charles Darwin suggested the latter in *The Descent of Man* (1871): "The perception, if not the enjoyment, of musical cadences and of rhythm is probably common to all animals and no doubt depends on the common physiological nature of their nervous systems."[1] He believed musicality must be a trait of *all* animals, a trait with a clearly demonstrable biological basis. Darwin's theory provides the underlying assumption of this book, in which I search for signs of "the perception, if not the enjoyment," of listening to music.

Music and Musicality

Scientists agree that music appears in all cultures, from the oldest civilizations of Africa, China, and the Middle East to the countless cultures of today's world. No culture has yet been found that does not have music. Music supports many social and cultural activities, from rituals and concerts to dance parties and funerals. It unites, consoles, and, simply, affords listening pleasure.

Yet some music researchers are skeptical about the biological foundations of musicality. In their opinion, every form of music in every culture is unique and is determined by human, social, and cultural conventions. If this way of thinking is true, then music and musicality have little to do with our biology.

The literature supporting this position usually restricts itself to music from Western culture, where music is generally the domain of professional musicians who have honed their skills with years of practice. But such a position does not do justice to the presence of music in all cultures and time periods. A broad range of research reveals that all people, not just highly trained musicians, have a predisposition for music in the form of musicality.[2]

Over the past few years, more and more systematic research has been conducted on the similarities and differences between music from around the world. Scientists are seeking "universals": formal structures and aspects of music that appear in every culture. If such universals are found, they can be used to support the idea that musicality is based on—or possibly constrained by—our biology.

But might there be limits on what is heard, experienced, appreciated, and passed on to future generations as music? The avant-garde composer Anton Webern (1883–1945) thought it was only a matter of time before postmen would be whistling his atonal melodies. Though postmen seem to be gradually disappearing, one might wonder whether the same applies to certain melodic and rhythmic patterns, in the sense that they are "whistled" down to the next generation. Observing all music cultures around the world, is it possible to say that some musical structures occur more frequently or much less frequently than others, or possibly never?

In 2015, a group of researchers from Canada, England, and Japan analyzed hundreds of music recordings from different musical traditions in North and South America; Europe; Africa; the Middle East; Southern, Eastern, and Southeast Asia; and Oceania. The study, based on classifying the music samples according to a long list of features drawn up by ethnomusicologists such as Alan Lomax[3] and Bruno Nettl,[4] revealed huge differences. But surprisingly—and this finding constituted the major contribution of this research—the analyses also demonstrated uniformity: specific structural features of music were found in nearly all the recordings. The melodies usually comprised a limited set of discrete pitches (seven or less). These pitches formed part of a scale that was divided into unequal and relatively small intervals. Most of the music also had a regular pulse (an isochronous beat), usually with two or three subdivisions, and a limited sequence of rhythmic patterns.[5]

In a certain sense, these universals are counterintuitive: one would expect to find considerably more variation, certainly based on the continuum of possibilities that tone and rhythm allow for in theory. For example, a scale can be subdivided in an infinite number of ways. It would appear that dividing this continuum into a limited number of discrete units—specific tones with a specific duration—has, cognitively speaking, a particular appeal for humans. A limited number of pitches and rhythms can be remembered easily but also combined in an endless number of ways. These characteristics ensure that the resulting melodies can be learned with relative ease, then passed on to subsequent generations. In addition, a regular beat facilitates anticipating the rhythm and tempo of the music, a prerequisite for being able to dance and make music together.

These findings offer exciting insights when it comes to analyzing the structure of music and the underlying similarities, but they tell us little about the biological foundations of musicality. This is because the method is indirect: the object of study here is music—the result of musicality—rather than musicality itself. It is also difficult to distinguish between the individual contributions of culture and biology. For example, it is not clear whether the division of a scale into small and unequal intervals in a particular music culture results from a widespread music theory doctrine or from a music perception ability or preference.

This book is therefore not about the structure of *music* but about the structure of *musicality*, musicality being a set of natural, spontaneously developing traits based on, or constrained by, our cognitive abilities (attention, memory, expectation) and our biological predisposition. Ultimately, I hope to be able to say something about the biological foundations of musicality and our innate musical ability.

Comparative Research

As with people, so the components of musicality have a genealogy that branches out across the globe and goes far back in time. Along with being a cultural artifact, musicality probably also has a biological basis. Unfortunately, since

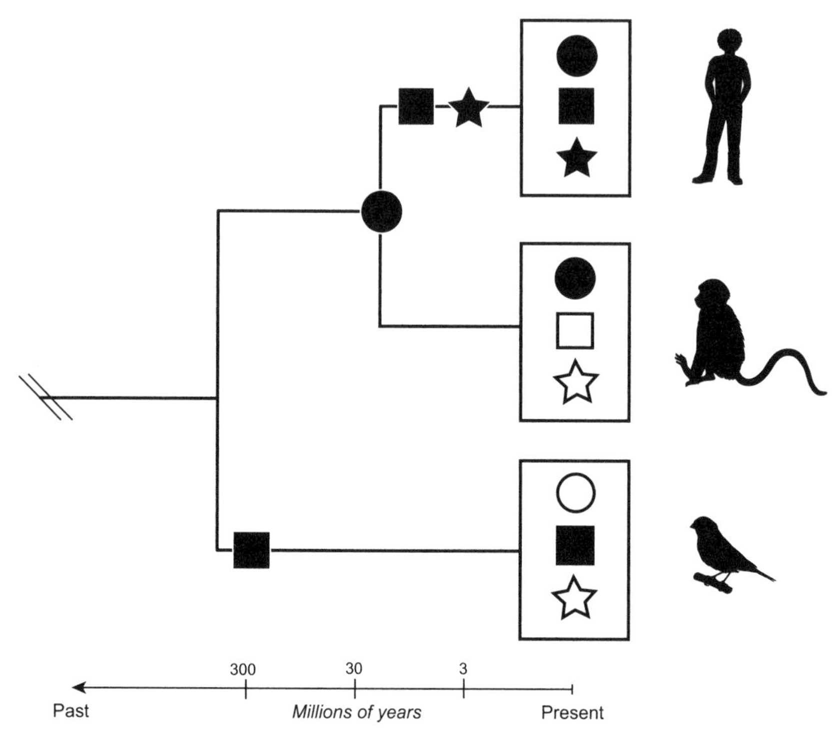

Figure P.1
Phylogenetic tree of musical traits that humans may share with rhesus macaques and zebra finches. Filled shapes represent a hypothetical musical trait (such as beat perception or relative pitch); open shapes indicate the absence of that trait. The position on the phylogenetic tree dates the possible evolutionary origin of such a trait.

neither musical sounds nor our musical brain fossilize, no physical traces of the history of musicality can be found. Comparative cognitive biological research, on the other hand, does allow for mapping out the genealogy of musicality. By studying whether related or unrelated animals share a certain musical trait, we can say something about their shared biological history (figure P.1). If two related species share a particular trait, their common ancestor would probably also have had that trait. This is how we date the origins of a specific musical characteristic. If two unrelated species share a characteristic, this may say something about the underlying mechanism and the natural selection pressure that may have led to that characteristic. Comparative cognitive biological research thus gives us a powerful method for learning more about the human musical ability, despite the absence of fossils from the distant past. When I began writing this book, I knew relatively little about the field of cognitive biology. However, together with more familiar methods from psychology and informatics (the cognitive sciences), it offered me new and intriguing perspectives on musicality that in turn led to surprising new insights.

My interest in biology began when I was working on *Musical Cognition: A Science of Listening*. In that book, based on then-recent research results, I substantiated the theory that we are all born with a predisposition for music, a predisposition that develops spontaneously and is further refined by listening to music. Nearly everyone possesses the musical skills essential to experiencing and appreciating music. Think, for example, of *relative pitch* (recognizing a melody separately from the exact pitch or tempo at which it is sung) and *beat perception* (hearing regularity in a varying rhythm). Even human newborns turn out to be sensitive to intonation (melody), rhythm, and the dynamics of the noise in their surroundings—the building blocks of music. Everything suggests that human biology is already primed for music at birth with respect to both the perception and enjoyment (to use Darwin's terms) of listening.[6]

Through these experiences, I became increasingly convinced of the notion that our musical ability—in particular, those musical skills so second nature to humans, such as relative pitch and beat perception—has a biological basis. These core components of musicality struck me as so trivial yet at the same time so fundamental that the question inevitably arose: Who, in fact, is "everyone"? Is everyone all humans? Or all animals? (Biologists generally distinguish between human and nonhuman animals. I make the same distinction in this book, but for the sake of readability, I refer to "humans and animals," even though humans are also animals.)

Darwin assumed that all vertebrate animals perceive and appreciate rhythm and melody simply because they have comparable nervous systems. He was therefore convinced that human musicality had a biological basis. He also suggested that sensitivity to music must be an extremely old trait, much older than sensitivity to language. In fact, he viewed musicality as the source of both music and language, and attributed its presence in humans and animals to the evolutionary mechanism of sexual selection. Darwin wrote: "With all those animals, namely insects, amphibians, and birds, the males of which during the season of courtship incessantly produce musical notes or mere rhythmical sounds, we must believe that the females are able to appreciate them, and are thus excited or charmed; otherwise the incessant efforts of the males and the complex structures often possessed exclusively by them would be useless."[7]

These two assumptions—the biological foundations of musicality and the idea of musicality as the evolutionary starting point of music and language—became the keys to the quest I describe in this book.

When Are You Deemed Musical?

After the publication of *Musical Cognition,* I was regularly sent YouTube videos of animals the senders considered to be more musical than themselves: an Amazonian songbird performing a bel canto song, a parrot belly dancing to Arabic music, a dog playing the piano, elephants moving slowly to the beat of a violinist playing Bach. Though fascinating to watch, the videos emphasize the—to us—familiar human perspective. We see and hear musical creatures in these animals. The singing by an Amazonian songbird readily sounds like music to our ears. Yet this says more about us than about the songbird. The question is whether a bird hears and experiences the same thing that we do. To better understand a bird's musicality, the question should really be: does a bird hear bird sounds as music?

This book is therefore not about whether a songbird or a gibbon or a humpback whale makes music—incidentally, the simple answer is yes, certainly to our ears—but rather about what you, as a human being or an animal, need to know or be able to do or feel in order to experience something as music. Reformulating the question in this way shifts the attention from music to musicality. The key components of musicality (relative pitch and beat perception), not the key components of music (such as melody and rhythm), then become the subject of study.

By now, it will be self-evident that we humans are musical (I refer readers in doubt to *Musical Cognition*). But the extent to which we share this ability with other animals was far from clear to me when I began writing this book. The question that kept presenting itself was: Is musicality something uniquely human? Or do we, in fact, share musicality with other animals on account of the "common physiological nature of [our] nervous systems," as Darwin suspected? I was determined to find out.

Personal Account

One way to map out the biological foundations of musicality is to conduct comparative research on animal species that share a specific musical trait with humans, a trait that is fundamental to experiencing music. It is useful to begin with "animal models" (animals we already know a lot about owing to research over many decades, and which therefore contribute to the understanding of a specific human function or trait). It is not laziness that prevented me from rushing off to the jungles of northeast Thailand to observe gibbons or from noting the nightingale's song in the Dutch dunes at daybreak. I am simply not a biologist. It was therefore vital that I quickly familiarize myself with biology's existing scientific methods and insights. To this end, I studied the relevant literature and engaged with biologists. Without their extensive knowledge and love of their discipline, I would most certainly have lost my way. The question I kept asking them and myself was: does the human musical ability have a biological basis?

Numerous biological research centers were kind enough to allow me to visit them. Each center had an expertise in the area of a specific "animal model." For research on the auditory system, *Macaca mulatta*, better known as the rhesus macaque, and *Taeniopygia guttata*, the zebra finch, are the animals of choice. Thanks to them, our knowledge about hearing, speech, and the brain in general continues to improve. Such research is also the reason why rhesus macaques and zebra finches play a prominent role in this book, with chimpanzees, cockatoos, and sea lions also making an occasional guest appearance.

In genetic terms, rhesus macaques are closely related to humans and therefore have similar brain structures. This species is thus a much-used animal model for neurobiological research on, for example, the cause and treatment of hearing and movement disorders in humans. Much of what we know about brain function can be attributed to biomedical research on these monkeys.

Zebra finches, on the other hand, are far removed from us genetically. Bird brains have a totally different structure from primate (ape and human) brains. Yet, despite the often radically different evolution of their brains, certain songbirds have traits comparable with those of human musicality. They can, for example, learn new songs. Through research on songbirds, we are learning more and more about the role of genetics, environment, and the evolution of their musical abilities.

My quest thus took me to the laboratories of Hugo Merchant at the Universidad Nacional Autónoma de México and Carel ten Cate at Leiden University in the Netherlands, where I familiarized myself with the practice of neurobiology and behavioral biology, respectively. I subsequently visited a marine mammal laboratory in Santa Cruz in the United States and a primate laboratory in Inuyama, Japan.

I feel privileged to have been able to look behind the scenes in all these laboratories and to have benefited from the knowledge and experience of their researchers. I am also grateful that the scientists considered my research question interesting enough to warrant music cognition research. Thanks to this interdisciplinary collaboration, I have come closer to finding an answer to my question.

Finally, this book is about science but it is not a scientific book. It is a personal account of my first steps into cognitive biology as a music researcher. Do not, therefore, expect a structured overview of available facts in support of a hypothesis (as in *Musical Cognition*). Rather, this book is an associatively assembled log of my exploration of a new research area, including all the unforeseen twists and turns, doubts, and oversights that are an inherent part of scientific research. It is a search for answers to the question of what makes us musical.

Amsterdam
March 2018

1 Shaved Ear

Querétaro, Mexico, October 31, 2011. You can see he knows something is about to happen. From inside his metal cage, Capi's large, black eyes follow his carer. Elsewhere in the fluorescent-lit room, a Mexican student by the name of Ramón Bartolo prepares for a new workday. He dons a white laboratory coat bearing the embroidered blue letters "UNAM Académico," hastily fastens the buttons, then pulls on two elbow-length leather gloves. The right-hand glove has seen better days: several shreds of leather are held together with duct tape.

Ramón claps his gloves together gently to get Capi's full attention. This is nothing new for the seven-year-old rhesus macaque: he is taken from his sleeping quarters every workday at the same time. The moment the sliding door of his cage opens, he grabs the gloved hand decisively and assuredly. Then, Ramón elegantly sweeps Capi down from his relatively high cage to his workstation.

The workstation resembles a high chair on wheels: it is custom-made from Plexiglas and has two aluminum cylinders as armrests. As usual, Capi himself will soon insert his arms into the cylinders. First, however, a black Plexiglas tabletop is pushed down over his head. He clings to it tightly almost immediately, as if to make himself feel more secure. Finally, Ramón slides a U-shaped Plexiglas sheet over Capi's long legs and small abdomen.

Capi peers around calmly, closely tracking his carer's every movement. As soon as he is properly settled in his chair, he is wheeled from his sleeping quarters into a corridor to be weighed. In a daily ritual, his weight is recorded on a chart hanging above the scales. The chart has four columns, one for each of the four rhesus macaques that live and work here.

The calendar hanging next to the chart features a color photograph of a different monkey species for every month. With his whitish-gray fur coat and large patches of green-gray hair, Capi unmistakably resembles the January photograph. The caption reads *Macaca mulatta*.

After weighing Capi, Ramón wheels him into the medical area. Capi releases the tabletop and drops his arms loosely behind him so that Ramón can administer his daily injection. With a small dose of the analgesic ketamine, Capi will undergo all the further preparations without a fuss. Unlike other days, today the hair will be shaved off his right ear.

By Invitation

I have been in Mexico for several days now. It is my first time in the country, and I am visiting the Institute of Neurobiology, part of the Universidad Nacional Autónoma de México (UNAM), in Querétaro, a small, affluent university city some three hundred kilometers northwest of Mexico City. Hugo Merchant, head of the UNAM primates laboratory, invited me here to investigate whether it is possible to perform the same listening experiments on rhesus macaques as I have previously performed on adult and newborn humans. Three months earlier, at a conference dinner in Leipzig, we had decided to devise an experiment to test beat perception in these monkeys. Although somewhat nervous about what lies ahead, I am also excited. Such a listening experiment may offer insight into the extent to which humans share beat perception with other primates, and the outcome may say something about our predisposition for music.

Beat perception—or, in the scientific jargon, beat induction[1]—can be defined as hearing regularity in a varying rhythm. Thanks to beat perception, we can identify whether the beat of music is regular, and whether it speeds up or slows down. Beat perception develops naturally in newborns and young children. Children move spontaneously to the beat of music, scarcely requiring instruction. Why is this? Are they already primed for music?

Some scientists believe that beat perception helped groups of early humans stay together. If humans want to dance, sing, make music, or even march together, they all need beat perception—the ability to "think" in the same tempo. Without it, singing and dancing with each other would not be possible. These are the kind of activities that bind, in other words, that are crucial for a group feeling. Primates may still engage in social grooming, but groups of humans probably quickly became far too large for such a time-consuming activity. This is where musicality and moving together to the sound of music offered an excellent solution, one that appears to have remained intact throughout evolution.[2]

Such an interpretation, however, is difficult to validate scientifically, simply because the brain does not fossilize and therefore leaves no physical traces. For this reason, I was eager to investigate whether beat perception also exists in a rudimentary form in our nearest animal cousins. Through comparative research, I hoped to learn more about a possible shared biological history. It was therefore a golden opportunity for me to be able to conduct the same research on both human *and* nonhuman primates. Though I had some doubts about the potential results, I had eagerly accepted Hugo Merchant's invitation.

After a number of lengthy discussions, it would have been easy to let Hugo and his colleagues conduct the research themselves, without traveling to Mexico myself. An online video link would have sufficed for sorting out any further details or questions. But I wanted to observe the workings of a neurobiological laboratory with my own eyes. Scientific journals often publish spectacular research results on the basic functioning of the cognitive, visual, and auditory systems based on neurobiological research. Think, for example, of the discovery of mirror neurons, the development of prostheses controllable by brain functions, or the treatment of serious human conditions such as deafness, polio, or Parkinson's disease.[3] For me, however, neurobiology was a completely new research area, and I was still uneasy about whether I wanted, or would be able, to bridge this gap. Though naturally I looked forward to observing the rhesus macaques firsthand and watching how they reacted to sound and music, I also wanted to be confident they were treated in a way I could feel comfortable with. After all, music research is not biomedical research.

Later that afternoon in Querétaro. Instructions for the animal carers at the Institute of Neurobiology are posted on a notice board on the wall of the rhesus macaques' sleeping quarters. The researchers' messages about the weekly feeding schedule, water rations, and exposure to light are written in black and green felt-tip pens and signed with an elegant *Gracias* (thank you). All this information is carefully recorded in tables and to-do lists.

In the adjacent control room, a rhesus macaque who has just begun his daily routine can be seen on a small black-and-white screen. His name is Yko. He is operating a joystick while simultaneously watching a screen and sucking on a straw. The grainy black-and-white image on the monitor in the control room reminds me of the film and photographic material of the first experimental spaceflights. As a child, I saw images of a rhesus macaque in a space suit buckled into a chair, passively awaiting what was to come. But Yko is not waiting: he is playing a computer game with gusto.

A second screen shows a close-up of the large, clear eyes and expressive eyebrows of the macaque at play. A small white cross on his black pupil confirms that the eye-tracking system is following the movement of his eyes effortlessly. His eyes dart back and forth in pursuit of the figures on the computer screen. These continually appear in different positions and different colors and sizes. Yko operates the joystick deftly, first toward one figure, then toward the other. He has learned how he is supposed to move the joystick: after hearing a short sound, in the direction of the blue circle; after hearing a long sound, in the direction of the orange circle.

Children play similar computer games to score points and reach higher levels. In this laboratory, the rhesus macaques do everything for a reward in the form of Tang, an orange drink popular among Mexican children. It is stored in a laboratory bottle high up on a shelf in a corner of the testing area. A precise amount is released at a constant rate via a dosing system and offered to Yko through a straw. The click of the dosing pump in the control room is heard whenever Yko responds correctly. After Yko gets the hang of the game, the dosing occurs about every ten seconds.

Capi, the rhesus macaque that will likely take part in our new listening experiment, is familiar with testing. He was two years old when the laboratory purchased him from a North American laboratory, and he is now seven. He was given his name, an abbreviation for *El Capitán* (the Captain), because he often holds his hand to his brow as if saluting. It turns out he has risen to the rank of alpha male in a group of four rhesus macaques that perform behavioral tasks here for three or four hours every day. Every morning, they are weighed and taken to the testing area; then, at the end of the afternoon, they are returned to their sleeping quarters. The light is switched off at night and on again in the morning.

Structure and regularity define the life of the rhesus macaques, as they do the life of all the primates in this laboratory. The workday for everyone lasts from 8:00 a.m. to 8:00 p.m. on weekdays, with a lunch break at about 3:00 p.m., and from 8:00 a.m. to about 2:00 p.m. on Saturdays. Sunday is the only free day. The two students who take turns picking me up at my hotel confided in me earlier that they find their work regimen too strict. It leaves them little time to see family and friends.

Today Capi will be shaved, because his coarse hair makes it difficult to perform an electroencephalogram (EEG). While a dab of conductive gel and an adhesive plaster are sufficient for humans, an electrode slips off a rhesus macaque at the slightest movement. Ramón uses a bright blue disposable razor to shave Capi's right ear, where the reference electrode will be attached. Slightly sedated by the ketamine, Capi undergoes the shaving passively. It is both an endearing and a distressing sight. His ear is being shaved because I want it, not he.

Today is also the first time I have been able to observe Capi at close quarters without him becoming agitated by my presence. Though I can see from his facial muscles that he is still drowsy, he follows every movement made around him. The soft white hairs on his ear are easily shaved off. After the coarser, green-yellow hair is removed from his scalp, only a downy white layer remains, thin enough to allow the EEG electrodes to be attached with medical tape.

The skull of a rhesus macaque is not only much smaller than I had imagined but also much flatter, rather like that of an adult cat. Hugo shows me a segment of skull that he has removed from another macaque, a piece of the crown the size of a small matchbox. A tiny segment of bone has to be removed for the measurement method commonly used in such laboratories. It involves inserting several electrodes into the brain using a catheter.

Sitting calmly in his chair, Capi, with his bony little head, reminds me of Paco, my onetime Siamese cat. I have to stop myself from giving him a pat on the head.

As expected, the effect of the analgesic wears off quickly. Capi's big eyes are as alert as ever, and his mouth, with its sharp eyeteeth, has firmed up again. It is no luxury that Ramón wears leather gloves when leading Capi from his cage to his chair.

I sit at eye level next to Capi and address him in a friendly tone, the way I would a cat or a small child. It quickly becomes apparent that this is not a good idea. He responds with anxious growling noises. Startled, I take a few steps back. Ramón laughs at my reaction, then reassures me it is my direct stare, not my friendly words, that has upset Capi.

Index for Beat Perception

Do primates have beat perception? I am often asked this question during lectures when I say that all humans are musical and that musicality is already active very early on in a human life. After all, as many people believe, is the heartbeat not the basis for rhythm in music and thus also for beat perception? If primate infants hear their mother's heartbeat in the womb as human infants do, then they too must have beat perception. This seems a logical line of thought.

Since a study that I completed in 2009, together with the Hungarian neuropsychologist István Winkler, we have known that human newborns have beat perception. In that study we demonstrated that infants who are only two or three days old already have it, in the sense that their brains are "surprised" when the first beat (the downbeat) of a varying drum rhythm is omitted. We reached that conclusion by measuring electrical activity with an EEG. Using several electrodes carefully attached to the newborns' heads, we observed small "potential differences," reflections of brain activity in response to sounds or, indeed, the *absence* of reflections precisely when one might have expected them.

Listening to music entails meeting expectations—a rhythm or melody you can tap the beat to or sing along with—as well as violating those expectations in the form of a surprise turn (a deviation or variation in the rhythm or melody). A violation of the expectation by a surprise turn is registered in the measured brain signal as a *mismatch negativity* (MMN), a distinct negative peak immediately (within about 150 milliseconds) following the unexpected turn. The MMN is thus a much-used index for measuring a subject's expectation while he or she is listening. Generally speaking, the more unexpected the sound, the earlier (and larger) the peak in the brain signal.[4]

Figure 1.1 shows a so-called *event-related potential* (ERP). An ERP is a time-locked fragment of an EEG signal that occurs immediately after a sound (an event) and, with the help of a computer, is averaged to eliminate irrelevant brain signals. The dashed line represents the ERP signal after an expected tone (the *standard*); the solid line represents the ERP signal after an unexpected tone (the *deviant*). The peak in the signal generated in response to the unexpected event is the MMN.

Using this electrophysiological technique, a listener's metrical expectation, or "beat perception," can be measured by observing the brain's

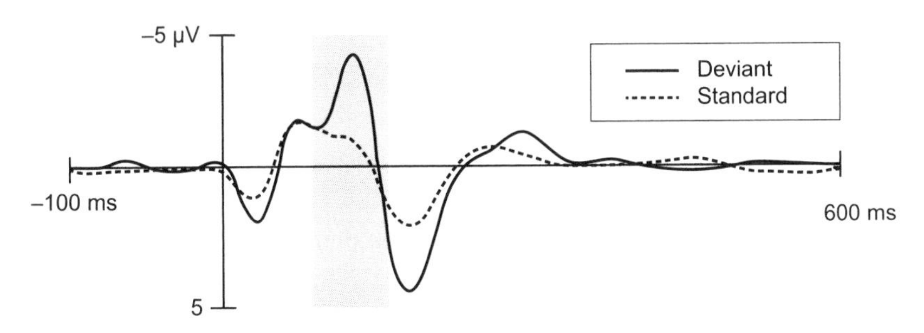

Figure 1.1
The average electrical activity measured on the surface of a listener's skull, expressed as an event-related potential (ERP), in response to expected (standard) and unexpected (deviant or oddball) sounds.

reaction to the unexpected omission of notes at different positions in a repeating rhythm. Occasionally the listener does not register the omission of the note; the silence goes unnoticed. But at certain positions in the rhythm, such a silence jumps out at the listener like a "very loud nothing." At that point, the rhythm falters. It is "syncopated," which is not the case for the equally silent intervals at less important metrical positions. That a particular rest jumps out at the listener—a so-called *loud rest*—whereas another physically identical rest does not, serves as an index by which to measure whether someone has a specific metrical expectation. We therefore used the MMN as an index for beat perception, even when the newborns were sleeping.[5]

We performed these experiments to test the beat perception of both adults (musicians as well as listeners who were not musically trained) and newborns. With the permission of, and in the presence of, the parents, we measured the reactions of dozens of newborns as they listened to a varying drum rhythm while asleep; the rhythm was played so softly in their headphones that it didn't wake them. As with the adult listeners, we observed the same characteristic peak in the infants' brain signal: an MMN. It turned out that the infants, who were only a few days old, already possessed beat perception.[6]

This discovery (along with findings of fellow researchers reported in several publications in the same year, about which more later) was responsible for an important shift in my research. Together, all the results raised the question of whether or not beat perception is based on a biological predisposition, and if so, whether our animal cousins have the same predisposition as we do. If beat perception is such a fundamental trait, then other primates must have it too.

The Rhythm Paradox

In *The Descent of Man*, Charles Darwin wrote that humans probably share the perception of rhythm and regularity with all mobile animals. A bird, for example, flies with a regular wingbeat to stay aloft, and most animals walk with a particular rhythm, without which they would fall or stumble. One would thus expect beat perception to be present in all animals that need to coordinate their body parts. Logically, their motor system must be rhythmic and regular.

Though Darwin used words like "rhythm" and "cadence" rather than the term "beat perception" to highlight the rhythmic behavior of animals—from shuffling to walking, and from swimming to flying—the facts available to date turn out to contradict Darwin's assumption.

In Hugo Merchant's laboratory, for instance, they tried to teach rhesus macaques to move a joystick regularly and synchronously to the ticks of a metronome. The experiment failed. After months of trying, the movement of the macaques was still no more than a reaction to individual ticks.[7]

Whereas humans begin making such a regular synchronized movement *slightly ahead* of the tick so as to be on time, the macaques only reacted simultaneously with the tick and so were always late. The study revealed that rhesus macaques react, while humans generally anticipate. Rhesus macaques appear to be unable to predict the regular tick of a metronome. For them, each tick is equally unexpected.

An alternative explanation might be that rhesus macaques do hear if the music speeds up or slows down and therefore do hear the regularity, but are unable to *anticipate* the change because their motor system does not allow it, just as some people cannot dance in time to music. It is therefore important to distinguish between "hearing regularity" (perception) and "producing a regular rhythm" (production). Darwin didn't make this distinction.

The contrast between Darwin's assumption and the facts available thus far is what we might call a "rhythm paradox": while regular and rhythmic movement is a widespread phenomenon in the animal world, it appears that the ability to move to the beat of the sound or music is limited to only a few animal species, including humans.[8]

The inability to make regular movements clearly does not imply the inability to hear the beat of music. Unfortunately, a behavioral experiment like the one Hugo Merchant performed with rhesus macaques did not allow for such a distinction. In an experiment where the movements of a test subject are measured, we never know exactly what causes the subject's behavior. Was the sloppy timing in the macaque's maneuvering of the joystick the result of an unsteady hand or an inability to hear the tick of the metronome clearly?

To circumvent this problem, I wanted to use the same method with the rhesus macaques that we had used earlier with newborns. As with newborns, it is difficult to determine, based on their behavior or movements, what rhesus macaques hear. The idea was to measure the brain activity of a rhesus macaque as it listened to rhythms that occasionally violated both the expectation and the beat perception. To me, it seemed so utterly simple to apply the same technique that we had used with newborns—namely, attaching five electrodes to the skull and letting the subjects listen to the rhythms—that I couldn't fathom why it hadn't already been done long ago with the macaques. Together with several students, I sifted through the neurophysiological literature, in which rhesus macaques appear all too often. But in all cases, the researchers were still using invasive techniques involving electrodes being implanted in the brain.

To my amazement, we found virtually no studies measuring EEGs in rhesus macaques using noninvasive techniques like the ones we had used on adults and newborns. One would think that this technique, which today is a standard diagnostic tool for different brain functions and brain disorders such as epilepsy, would have been developed with the help of rhesus macaques.

I raised the question with a number of neurophysiologists. The answer I got was simple: it is not possible to measure brain activity in rhesus macaques using an EEG. Their skull, hair, and head muscles are simply too thick to allow this type of electrophysiological measurement. Moreover, their constant moving and chewing would make any measurement of brain activity virtually unusable. For this reason, neurophysiologists prefer an invasive method involving the implantation of electrodes in the brain: the brain activity of a specific area can then be measured directly and without interference. Perhaps the higher medical goal justifies the means, but research into beat perception has no medical urgency, and I wanted to measure without having to cut—or having someone else cut—into the body of the rhesus macaque.

Frankfurt, May 18, 2011. Six months before my visit to Mexico, I attend the Frankfurt Institute for Advanced Studies (FIAS) as a guest for one week, together with about thirty international colleagues from a range of disciplines. Our goal is to discuss language, music, and the brain. During this interdisciplinary conference, I find the key I have been seeking for months.

Fierce debates about the differences and similarities between language and music dominate the morning sessions. The first few days are quite difficult because of the major differences in scientific background and particularly style, with the assertive researchers from the United States on the one hand and the more reserved, cautiously hypothesizing, and sometimes even shy Europeans on the other. Though I am a little jealous of the participants who express their point of view confidently and with no apparent doubt, I am initially quite reserved and focus on asking questions.

Though I have forgotten the exact details, I think it must have been one of the behavioral biologists in the group who, during a heated discussion, mentioned the name of a researcher and a date in response to a remark from a neurophysiologist. This often happens, particularly during exchanges of opinions among scientists from different disciplines who do not know each other very well yet. Citing the name and date of a publication is a way to impress the other and is part of normal disciplinary territorial behavior. You reveal that you have knowledge at your disposal, knowledge the other person clearly lacks because he or she reads different scientific journals. It amounts to staking out your position in terms of knowledge. But, of course, there is no such thing as an omniscient scientist. This is one of the reasons interdisciplinary workshops are essential for the advancement of science.[9]

The neurophysiologist's assertive statement referred to the impossibility of performing an EEG on a primate: to date, no one had ever done it successfully. The head muscles of primates are simply too active, and the electrical signals they produce overshadow the relevant but infinitely smaller brain signals in the EEG. Provoked by the assertiveness of the neurophysiologist's remark, however, the behavioral biologist who cited the reference gave me precisely the information I had been seeking for some time.

I type in "Ueno 2008 EEG" on my tablet and instantly find the relevant article. I go straight to the part of the text reporting the test results. The figure that summarizes the results is unequivocal: it is definitely possible to perform an EEG on primates and to measure an ERP.

The key player in this study by Ari Ueno and his colleagues was Mizuki, a nine-year-old female chimpanzee who had grown up in the Great Ape Research Institute in Okayama, Japan. Mizuki had often participated in behavioral experiments in the past. In this particular study, scientists had attached several surface electrodes to her skull to measure her EEG as she performed a task. This involved simply listening to a sequence of tones while doing nothing.[10]

It took the researchers about six months to train Mizuki to become accustomed to the procedure and the attaching of the electrodes. This is normal. Hugo Merchant's rhesus macaques, like test animals in other laboratories, are trained to stay calm and relaxed while undergoing experiments. Just as a horse must get used to a saddle, a bit, and a person sitting on its back, so rhesus macaques must become accustomed to a specific wake–sleep rhythm, work area, reward, and the method itself before the research can truly get under way.

The Japanese researchers' method of working was identical to the "oddball" paradigm method that we had used with adult humans and newborns. This method measures the brain signal in response to an unexpected event (see figure 1.1). In the case of our newborns, the event was unexpected silences (loud rests); in Mizuki's case, it was sounds with an unexpected pitch.

Most of the time, Mizuki was exposed to the same tone (the standard), but occasionally this was replaced by a slightly higher tone (the deviant or oddball). Think, for example, of a clock that ticks regularly but occasionally deviates. The deviant tone violates the expectation and is therefore more noticeable.

Mizuki was also exposed to the opposite: a higher standard occasionally replaced by a lower variant. This was to ensure that the deviation from the acoustic regularity, rather than the frequency of the tone, would elicit the slight peak in the brain signal. Sure enough, in both cases, a negative peak occurred in the ERP, thus closely—and surprisingly—resembling the way the human brain responds to an unexpected sound.

The Ueno study dispelled many of my doubts. Though the experiment tested Mizuki's reaction to unexpected tones rather than beat perception, the method could clearly be used to measure expectations in nonhuman primates' listening behavior without resorting to invasive methods. The ability to measure an ERP made it possible for me to compare, in a relatively simple way, the nature of the "musical" listening of human and nonhuman primates, and to examine the extent to which melodic and rhythmic expectations play a role in primates' listening.

Seating Plan

Leipzig, July 15, 2011. Three and a half months before my trip to Mexico, I visit the Max Planck Institute for Human Cognitive and Brain Sciences in Leipzig, where the international Rhythm Production and Perception Workshop (RPPW) is taking place. This relatively small, biannual event brings together about fifty researchers from diverse disciplines to discuss rhythm and different rhythmic activities, such as rowing, golfing, walking, talking, and, of course, making or listening to music.

The customary conference dinner is scheduled for halfway through the week. Unlike weddings or other official events, such dinners rarely have a seating plan. Rather than listening to the familiar stories recounted by my colleagues, I opt for chance. A fresh perspective on one's research questions can be extremely informative. I sit down at the head of a still-empty table. One of the first people to join me is Hugo Merchant. It is the first time our paths cross.

Earlier that day, Hugo had given a lecture on his recent neurobiological research. His work had revealed that different sorts of "timers" can be localized in the brain, in this case, in the brains of rhesus macaques, which had served as the model for human brains in his study. In rhesus macaques, specific brain cells in the supplementary motor area (SMA) or cortex, which forms part of the motor system, are active during a countdown to the moment when something will happen (think, for example, of being able to press a button on time by counting down), whereas other brain cells measure the elapsed time after something has happened (think of estimating the time span between two events). One brain cell population encodes the remaining time, while the other encodes the time elapsed during a motor movement. Both synchronizing with a sound and maintaining the tempo appear to depend on a coordinated interaction between the two timers.[11] Last but not least, both are vital for dancing to or making music. If tempo cannot be synchronized and maintained, it is impossible to dance to the beat of music or make music together.

Hugo and I talk at length about the interpretation and various consequences of the results, such as estimating time, the notion of a mental clock (either centrally located or distributed throughout the brain), and the difference between recognizing rhythms (sense of rhythm) and recognizing regularity (beat perception). Clearly a passionate scientist, Hugo's eyes sparkle as we talk. He is eager to know how time and a sense of time are effectuated in the brain.

I tell Hugo about an article I had read a few weeks earlier in which Japanese researchers claimed to be able to measure ERPs in chimpanzees. Hugo reacts differently from most of the neurophysiologists I have spoken to. Unlike them, he believes it *would* be worthwhile to test the method.

We spend the rest of the evening discussing various technical details, the pros and cons of the method, the young rhesus macaque Capi that Hugo is raising, and related research currently being conducted in the area of musical cognition, like the studies on beat perception in human adults and newborns.

At the end of the evening, I get straight to the point. Does Hugo think it is possible to measure an MMN in rhesus macaques? "That is an empirical question," he replies provocatively. "You have to do it to find the answer." To which he adds decisively, "So let's do it!"

I hadn't anticipated such a reply. Overwhelmed by Hugo's offer, I leave the table briefly. In the hallway to the washroom, I can't resist kicking my heels in the air, a typical 1970s disco move unconsciously elicited by the music emanating from the restaurant.

Within three weeks, Hugo Merchant and his team have all the software and hardware in place to be able to make the first trial measurements. This is fast, particularly for a laboratory that doesn't normally do EEG testing. Special equipment has been purchased and numerous programs written to test and calibrate the equipment. A separate electrical circuit has also been installed—identifiable by its orange power strips—to ensure there is no interference from the other equipment in the laboratory.

In the meantime, I am working with colleagues in Amsterdam to design a series of listening experiments that will allow us to explore the new research area. We use the same sounds as in earlier experiments, but this time, we add several intermediary steps.

I am keen to perform the listening experiment in three phases, advancing step by step toward the ultimate goal. After all, the skeptics might well be right that MMNs cannot be measured in rhesus macaques. I had repeatedly been warned that every movement of the rhesus macaque would interfere with the measurements.

I think it wise to begin by repeating the experiment that had been conducted earlier (and for the first time) with the chimpanzee Mizuki. Clearly that was not a study of beat perception but a test to see whether an MMN could be measured in response to an unexpected sound. Our first experiment will therefore be crucial, because if we cannot detect any MMN, the other listening experiments will be pointless.

If we *are* successful in measuring an MMN, we will then perform two other listening experiments. In the second experiment, we will investigate whether we can measure an MMN in response to a single unexpected silence (a loud rest) in a regularly sounding rhythm. This experiment will also have to succeed before we can embark on the final one, which is our primary focus: measuring beat perception in rhesus macaques while they listen to a varying drum rhythm.

We thus have a three-stage rocket ready to go: a series of experiments each of whose outcomes is crucial for the subsequent experiment, and the final experiment that will provide the desired insight. Will the brains of the rhesus macaques be just as surprised as those of human adults and newborns when listening to a faltering (syncopated) rhythm?

I am eager to know, but something still bothers me. Is it ethical to perform involuntary tests on nonhuman primates? I use my first few days in the Mexican laboratory to find my answer and come to a decision.

2 Mirroring

Stresa, Italy, July 23, 2005. Giacomo Rizzolatti of the University of Parma delivers the opening lecture this morning at a major international conference in Stresa on Lake Maggiore. With an Italian flair for drama, he shows video clips of a rhesus macaque participating in an experiment in his laboratory. The electrochemical changes in an active brain cell, measured live in the brain of a rhesus macaque while it is awake, are converted into sound. The audience hears fast clicking sounds whenever the macaque grasps a piece of apple. But Rizzolatti's surprise discovery—and the one that established his international reputation—was that that same fast clicking was elicited by the same brain cell when the macaque saw the person performing the experiment also grasp a piece of apple. The rhesus macaque's brain cell appeared to "mirror" the observed action.

The discovery was unexpected, as it was commonly thought at the time that sensory events (such as observing an object or action) and motor activities (such as grasping an object) involved distinct functions originating in different brain cells or areas of the brain. Rizzolatti shows a series of graphs refuting this interpretation. "The images speak for themselves," he declares proudly and confidently. "We don't need statistics." He goes on to talk enthusiastically about the various implications of these results, such as a better understanding of empathy, imitative behavior, and the role of mirror neurons in autism.[1]

During my visit to Hugo Merchant's laboratory in 2011, my thoughts return to this lecture and to the uncomfortable feeling it had left me with. After a few days in the Mexican laboratory, though, I am confident that the rhesus macaques here get more attention and care than cows, pigs, or chickens do on factory farms.

At the end of the week, Hugo invites me and several of his colleagues to his favorite restaurant, not far from the UNAM campus. During the meal, I ask Hugo and my fellow diners, mostly neurophysiologists working with rhesus macaques, a wide range of questions. How important is sound to rhesus macaques? What are their eating and sleeping habits? And why is this particular species of monkey so prominent in neurobiological literature? As the evening progresses, I get a clearer picture of the rhesus macaque, which, apart from differences in the brain, appears to have a great deal in common with humans, including socially.

Old World Primates

In addition to the oft-cited example of Darwin's finches, the evolutionary history of rhesus macaques is often used in textbooks to illustrate the concept of "adaptive radiation." This is a process in which organisms colonize a new environment, adapt to the local habitat, and subsequently develop into a variety of new species. Darwin's finches and rhesus macaques are thus attractive subjects for evolutionary biologists studying the interaction between organisms and their environment.

The ancestors of the rhesus macaque originated in Africa, where, at about the same time as our human ancestors—some twenty-three million years ago—they split off from other nonhuman primates. After migrating from North Africa to Europe, when these continents were still geographically close to each other—hence the name "Old World primates"—they spread to Asia, where new species emerged as they adapted to their new environments: from tropical rain forest to dry desert-like areas, and from mountainous terrain to marshes. Today, apart from humans, the rhesus macaque and its close relatives—the Japanese macaque, the Taiwanese macaque, and the crab-eating macaque—are the most widely distributed primates on Earth.

The rhesus macaque is also one of the most commonly used animal species in biomedical research. It has served as an animal model for decades because it is genetically close enough to humans to serve as a model for the working of primate brains in general. At the same time, it is distant enough from both humans and the great apes (chimpanzees, gorillas, bonobos, and orangutans) to allow medical procedures to be performed on it ethically that would not be permitted on healthy humans.

Most medical universities have a number of rhesus macaques hidden away in a laboratory somewhere. These monkeys appear to need relatively little care and attention and thrive in nearly all conditions, whether natural or artificial. Moreover, unlike other wild animals, which quickly tend to become lethargic and depressed in captivity, they generally remain alert and active.

As a rule, rhesus macaques live on a diet of water and dry food similar to dog biscuits. In Hugo Merchant's laboratory, there are always dog biscuits lying around untouched in the corners of the cages. But while every morsel of dry food looks and smells the same, many of them are rejected by the macaques. The primatologist Dario Maestripieri suspects that rhesus macaques carefully inspect, sniff, and touch each morsel to demonstrate to their environment that not just anything will do. The compliance of rhesus macaques appears to have its limits.[2]

Apart from the fact that rhesus macaques are genetically so close to humans, there is also a practical reason why this species in particular, rather than other primate species, populates neurobiological laboratories. At the beginning of the last century, rhesus macaques were readily available in India. After the results of the first experiments were published in the scientific literature, almost as a matter of course they became a much-used animal model in primate research. Even when India stopped exporting rhesus macaques at the end of the 1970s, they continued to be an important animal model. Today they are mostly imported from China, Nepal, and Puerto Rico, where they are bred in special centers.

In short, thanks to a fortuitous choice and the characteristic conservatism of the scientific community—which attempts, through small, precise steps, to map out all the nuances of a specific phenomenon—we are deeply indebted to rhesus macaques for their contribution to our medical knowledge.

Querétaro, November 2, 2011. I have resolved to observe everything that happens in Hugo's laboratory firsthand, including the less-agreeable things. That is why I am in the operating room today. Preparations are under way for an experiment that Hugo Merchant and his group have been busy with for weeks.

Every other day, the small tubes of surgical steel embedded in the skull of the rhesus macaque taking part in one of the neurobiological experiments are cleaned. On this particular day, the screw top of one of the tubes is removed. Hugo asks if I want to peer through the surgical microscope. Though hesitant, I look anyway and see a small piece of exposed brain, a grayish substance with tiny red blood vessels pulsating gently to the rhythm of the mildly sedated macaque's heartbeat. I estimate the rate of the heart to be about 100 beats per minute (BPM), an average tempo for music but relatively fast compared with the human heartbeat, the resting rate of which is about 70 BPM.

Hugo points to the supplementary motor area (SMA) or cortex, where he is taking the measurements: this area of the brain forms part of the network of the motor cortico-basal ganglia-thalamo-cortical (mCBGT) circuit, which, he explains, is involved in the planning and timing of movements, among other things. His current experiments aim to offer better insight into both the mCBGT network (which both humans and rhesus macaques possess) and the connections between the motor cortex and the auditory cortex, connections that appear to be less developed in rhesus macaques than in humans. This research is important for advancing our understanding of all kinds of human disorders, including Parkinson's disease.

In the meantime, the rhesus macaque is sitting unperturbed in his primate chair. He probably has no idea at all of what is taking place up there in his head. Like people with Parkinson's disease or epilepsy during an operation, he doesn't feel a thing. While he may suffer from a lack of freedom of movement, a vigorous scratch under the armpit or a good stroll or climb is out of the question at this moment.

After the tubes are cleaned and disinfected, a number of electrodes are implanted in the macaque's brain using a state-of-the-art device. In minute steps, micrometer by micrometer, each electrode is lowered through the metal tube using a remotely controlled electromotor. Any activity of a group of brain cells picked up by the electrode during its slow journey downward is heard over the loudspeaker as a click.

Juan Méndez, the student operating the device, is an expert by now. Depending on the frequency of the clicks, he can identify which is a good, active cell to measure that day. He is looking for a brain cell that "fires" when the rhesus macaque sees a moving object on the screen. For Juan, this means implanting electrodes every day, calibrating, listening intently in the hope of finding a relevant and active cell, and subsequently measuring the brain activity of the rhesus macaque for hours on end as he plays with the joystick.

During the lunch break, Juan shows me videos on his laptop of his vacation in India. A constantly moving camera closely tracks every movement of his eye, registering endless images of playing, climbing, and tumbling rhesus macaques that have made their home in a derelict temple complex. With their incessant screeching and vast numbers, entire troops of brazen rhesus macaques indisputably call the shots here.

I try to imagine what a strange experience it must be for someone who has spent most of his scientific career working with rhesus macaques in captivity to suddenly encounter them in the wild for the first time.

Visual Report

Hugo gives me free rein to explore his laboratory, question his coworkers, and even film. He is proud of his lab. A Dutch broadcasting station has lent me a small digital camera with which to photograph the preparations for the experiment. The images may eventually be used in a documentary on rhythm scheduled for broadcast in 2012.

Playing the role of photographer has a distinct effect on how I view my surroundings. The camera gives me an excuse to observe and take a position of distance. Though my role is primarily to guide, with the camera I can occasionally adopt a more inquisitive pose. Hopefully, this will allow me to become better acquainted with my new colleagues and their ways of working.

But the filming also captures the day's activities. I wake early because of jet lag and use the time to edit the previous day's images on my laptop and turn them into short visual reports: my first impressions of Mexico; the university's brightly painted red concrete buildings; the warm wind that sweeps over the campus, rocking the stately palm trees; the laboratory full of racks stocked with mostly homemade equipment; the staff members' coffee corner; the rinsing basin, with a packet of Tang, a banana, other pieces of fruit, and a head of lettuce for the rhesus macaques on the adjacent counter; the calendar featuring the primate portraits; and, of course, the rhesus macaques themselves, whom I see now for the first time at close proximity and outside their cages.

Back in my hotel room, the video images of the first pilot experiments look dramatic on my laptop screen compared with the video clips that I saw six years ago during Giacomo Rizzolatti's lecture. But this time I understand what

I am seeing. Watching the same video clips over and over again is somehow reassuring.

After working in the laboratory for several days and observing firsthand how the rhesus macaques live and are treated, I decide to accept responsibility for having one macaque listen involuntarily to my rhythms. I am finally ready to take the plunge.

Querétaro, November 3, 2011. Capi is transported to the room where, during the past few days, we did test runs of the listening experiment with carer Ramón as the subject. We had wanted to ensure that the measuring equipment and analysis software were working properly. The space is almost entirely filled with an aluminum structure to which all kinds of equipment can be attached in any position and at any height. The chair on which Capi sits rolls easily into position. Ramón uses a felt-tip pen to mark five dots on Capi's flat skull, indicating where the electrodes are to be attached. A cluster of colored cables for the EEG hangs from a beam. Each cable has a clear label indicating the codes corresponding to the different places on the skull (figure 2.1).

Luis Prado, the laboratory's engineer, hands over the electrodes, each of which has been dabbed with conductive gel. Hugo and Ramón attach

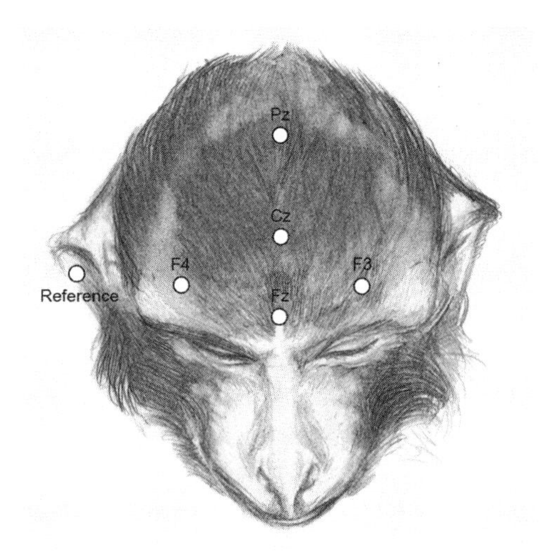

Figure 2.1
The skull of a rhesus macaque, showing the positions where the electrodes will be attached. (Illustration by Roos Holleman.)

them with generous amounts of medical tape. The end result is a rhesus macaque with a little cap of white tape on his head. Capi is stoical about his unusual new headgear, but the electrode attached to his right ear irritates him. He only stops touching it after the third attempt to attach it.

The listening experiment begins. We dim the lights in the room. The rhythms we so carefully devised for the listening experiment can be heard softly in the background. On the black-and-white monitor in the control room, I notice that Capi occasionally falls asleep. His head droops forward slightly, and though his eyes are still open, the gently undulating wave of the EEG confirms he is dozing. This is the moment when we measure his brain activity in response to the rhythms. Capi will participate in these dozing-listening sessions every day for the next two weeks except on Sundays.

Normally, the rhesus macaques in this laboratory take part in an experiment every afternoon. This might entail following a moving dot on a screen with a joystick while their brain activity is measured, after which they are given a reward in the form of a drop of orange juice. In our listening experiment, every movement causes a disruption in the specific brain signal that we are interested in. Our challenge, therefore, is to find a task in which the rhesus macaque moves very little and drinks, eats, or chews impassively.

As a result, we decide to perform the experiment in exactly the same way as we had with the human newborns in our study in the Netherlands. There is, after all, a comparable problem. When a baby turns its head or moves one of its eyes or hands, this produces "artifacts," or disruptions in the brain signal being measured. We found we were able to take measurements most effectively when the newborns were sleeping. We let them listen to rhythms during their normal sleep, and their brains responded to these rhythms without disruption. We therefore use the same setup for Capi. In this way, I hope to be able to forestall the cautionary words and critiques of my neurobiological colleagues. Over the course of several days, we alter Capi's day-and-night rhythm so that by the afternoon, he is very sleepy. As a result, he mostly sleeps or dozes during the listening experiment and barely moves.

Querétaro, November 5, 2011. After a few days of performing pilot experiments, we have our first result. Gábor Háden, my colleague in Amsterdam and an electrophysiological expert, has used the past two days (when it was nighttime in Mexico) to perform the first analyses. Following the repeated exchange of the measurement data and an online video discussion about the resulting graphs, we have an unequivocal answer: with the current design here in Mexico, we can definitely measure MMNs.

I am pleased and particularly relieved that, contrary to the predictions of my skeptical colleagues, we can now begin the actual listening experiment. Hugo and his team will continue to take measurements during the coming weeks. After working in the laboratory for more than a week, it is time for me to return to the Netherlands.

On the flight home, I read the stack of articles that Hugo has given me about recent research on the neurology of time perception in rhesus macaques. I also watch the video clips I had assembled during the week, and make a preliminary selection for the film director who loaned me the camera, this time with the eyes of a television viewer. I try to imagine how a viewer might interpret the images: Capi in his primate chair with five electrodes attached to his head, itself covered with a cap of medical tape; Hugo and his coworkers in their white lab coats; the different screens in the control room; and the shaving of Capi's ear. The images can be viewed in different ways, that much is clear. What remains in the end, though, is a feeling of excitement. What will the test measurements of the next few weeks reveal? Will rhesus macaques have beat perception like humans or not?

3 Beat Deaf

Amsterdam, November 11, 2011. It is evening now as I cycle to the University of Amsterdam's brand-new science park on the city's eastern outskirts. In a small lecture hall still smelling of fresh paint, mobile equipment has been set up to allow for a live video link with the other side of the Atlantic—specifically, with the International Laboratory for Brain, Music and Sound Research, known as the BRAMS institute, in Montreal, Canada. The team of the Dutch broadcasting station that loaned me the camera for my trip to Mexico is making a documentary in Montreal about "the man without rhythm." They were invited there by Isabelle Peretz, director of the BRAMS institute and an expert in the area of amusia (the absence of musicality).

Amusia is a condition characterized primarily by tone deafness and beat deafness. The first is the most common disorder: 2 to 4 percent of people in the West suffer from tone deafness to a greater or lesser degree. These people cannot detect when a wrong note is played in a familiar melody or when two melodies differ. In the worst case, music is an irritating racket best avoided.[1]

The second disorder, beat deafness, is even less prevalent. Over the past few years, the BRAMS institute has found only one person who could be diagnosed with this form of amusia. Today I am going to meet him "in person" via a live video link. His name is Mathieu.

On the wide-screen monitor, I see several technicians enter the studio space. The Dutch film director is sitting in the middle, savoring his sandwich while the light is adjusted. To keep the conversation spontaneous, we have discussed nothing in advance. The only thing we agreed on beforehand is the "reverse shots," which the assistant director, who stayed behind in Amsterdam, will make with a small camera. When I ask the director in the Montreal studio how the Dutch crew's visit has gone thus far, he tells me they went to a discotheque the evening before to film Mathieu on the dance floor. Mathieu had been visibly uncomfortable and appeared mostly to imitate the other dancers.

I am looking forward to the conversation with the man without rhythm. A surprising article about him revealed that he couldn't hear any difference between a march and a waltz. According to Isabelle Peretz, Mathieu is the first scientifically documented case of a beat-deaf individual.[2]

Dancing without rhythm is not uncommon. We all know someone who moves on the dance floor as if dancing to another tune. When Mathieu listens to a variety of rhythms in the laboratory and is asked to tap the beat, however, it turns out that something unusual is going on.

To further investigate Mathieu's lack of beat perception, Pascale Lidji, one of the BRAMS team researchers, had conducted a series of experiments during the preceding weeks, including the EEG listening experiment that we had performed earlier in Amsterdam.

Around seven o'clock, Pascale and Mathieu enter the recording room. Pascale sits in one of the chairs and proudly waves a sheet of paper containing the first test results. Mathieu looks rather nervous.

A clever, good-natured twenty-three-year-old, he claims to have no problem with the fact that he has no beat perception. "That must be a disappointment to someone who's so focused on beat perception," he jokes. After all, he manages fine without it. Yes, years ago, at school, any offer he made to sing or play the guitar was quickly rejected. And during ballroom dancing, his partner always had to lead because Mathieu couldn't hear the first beat. Mathieu compares his impediment to stuttering: it is awkward, but a lot can be done to ameliorate it. During the past four years, because of all the experiments he has participated in, he has learned "how to listen and move better."

From her seat next to him, Pascale reacts emphatically: "Solutions are only possible if you know what causes the absence of beat perception. Otherwise it's just personal avoidance of the problem." Mathieu looks disappointed and adds: "Yes, my experience will never be the same as other people's."[3]

After a lively discussion lasting about twenty minutes, filming stops. Isabelle Peretz steps out of the shadows and sits down next to Pascale and

Mathieu. The quality of the sound and images is so good that it is as if we are sitting on opposite sides of the table, except that it is afternoon there and evening here.

We reflect again on the conversation, agreeing that what we both most want to know is which cognitive and neurobiological processes enable beat perception or, indeed, suppress it. If we can identify what is absent or what is being suppressed in Mathieu, we may be able to take a crucial step toward understanding which neural networks play a role in musicality.

The absence of beat perception is an intriguing phenomenon for Isabelle and me. We are both convinced that beat perception is a unique and fundamental trait of musicality. However, while Isabelle believes it is uniquely human, I suspect that our animal cousins have it too, simply because our brains have a comparable structure (homology). For now, though, we still lack evidence. To date, the results of listening experiments measuring beat perception in humans and nonhuman primates suggest the opposite. These findings would mean that Isabel is right.

At the end of the video session, I ask Pascale for the first results from the listening experiment with Mathieu. What I see surprises me. There appear to be no differences between Mathieu's brain signals and those of a control group, comprising about fifty regular listeners of a similar age and with a comparable education and musical exposure.

Is it possible that Mathieu's brain unconsciously recognizes regularity in music and that things only go wrong during the reproduction of it, the rhythmic moving or tapping along to the beat of the music? Or is beat perception more cognitively based, a regularity that the brain registers but the person who is beat-deaf cannot consciously access?

It is still too early to draw conclusions. To give greater credence to the results, it is vital that the BRAMS institute, which is highly experienced in tracking down people with amusia in its many variations, find more beat-deaf listeners.

Around eight o'clock, I say farewell to my fellow researchers on the other side of the ocean. I can scarcely imagine what it would be like to be tone- or beat-deaf. I think I would miss the pleasure of listening to music enormously. I am also pleased I have got to know Mathieu without having to fly back and forth to Canada. Excitedly, with loud music resounding through my earphones, I cycle home in the crisp evening air.

The Man without Rhythm

The anticipated television documentary is broadcast in the Netherlands several weeks after the live video session. The documentary paints a portrait of Mathieu that incorporates, along with an interview with Isabelle Peretz, recent film clips of Capi, glimpses of Hugo's laboratory, and shots of our earlier listening experiment with human newborns. Understandably, the voice-over heavily emphasizes Isabelle's and my differing opinions. However, though I am certainly in good company with Darwin on this, we simply do not know yet. We are still waiting for the results of the listening experiments with Capi.[4]

During the next few months, the BRAMS institute continues to work intensively on the EEG listening experiment. Isabelle finds another beat-deaf listener, Marjorie, who participates with Mathieu in a new series of listening and behavioral experiments designed to test their beat perception.[5]

The final results of the EEG experiment are not what I expected. A "loud rest" elicits as large an MMN (mismatch negativity) in Mathieu and Marjorie as it does in normal listeners, which means their brains *do* have beat perception.

However, there are also differences. The P300 in particular—a positive peak that usually follows the MMN after about 300 milliseconds—is different in Mathieu and Marjorie than in the control group. It is smaller than normal. The P300 is an ERP (event-related potential) component that indicates the extent to which a specific stimulus is perceived by a test subject. Mathieu's and Marjorie's brains detect the regularity in a drum rhythm (as the MMN revealed) but have reduced access to this information (as the P300 revealed). Beat deafness therefore appears to be a primarily cognitive phenomenon. Beat-deaf people appear to lack the ability to perceive regularity in music although, at the neurological level, their brains register it normally.[6]

Isabelle Peretz had already demonstrated something similar in tone-deaf listeners. In that group, too, the absence of a conscious perception appeared to be what made the difference. She suspects this absence is caused by the neural network connecting the right frontal lobe (inferior frontal gyrus, or IFG) with the auditory cortex (A1). In tone-deaf listeners, the recurrent processing of auditory information in the right frontal lobe is often disrupted.[7]

Something similar may also happen in the case of beat-deaf listeners like Mathieu and Marjorie. Specific connections between certain areas of the brain may be less developed, connections essential for beat perception. All of this makes beat- and tone-deaf listeners information-rich subjects when it comes to mapping out the neural networks that have evolved specifically for processing pitch and rhythm and therefore also for musicality.

Although electrophysiological methods allow us to measure extremely accurately the amount of time it takes for the brain to react to unexpected events, unfortunately an EEG is less suitable for determining the source of the signals and their precise location. Finding the source requires other imaging techniques, such as magnetoencephalography (MEG) or functional magnetic resonance imaging (fMRI).

In 2011, Isabelle, Hugo, and I still did not know exactly which neural networks played a key role in beat perception, despite our understanding that it wasn't so much specific areas of the brain that were responsible for a specific musical function as it was the connections between these areas. More and more researchers were probing "functional connectivity," in other words, how different areas of the brain fulfilled a certain function by interacting closely with each other. The concept of functional connectivity offered a new way of looking at the existing scientific literature.

Was it possible that some neural networks had a longer evolutionary history than others? Could we identify similarities and differences between the neural networks involved in melody and rhythm perception in humans and nonhuman primates? What new insights might the evolutionary biological perspective offer if one were to take into account the most recent brain research? Bearing in mind Darwin's idea that musicality may indeed be older than both music and language functions, I eagerly devoured the literature on the subject.

Origins of Musicality

Though most theories in evolutionary psychology contend that the capacity for music and musicality benefits from the neural networks that have evolved for language (as suggested, e.g., by Steven Pinker; see chap. 6, "Supernormal Stimulus," later in this book), in fact quite the opposite may be true. Might language not benefit to a large degree from the much older networks involved with musicality, specific portions of which were adapted and reused only much later in human evolution?

It is not uncommon for neural networks to be recycled during evolution. Reuse seems to occur mostly when the older network possesses most of the neural structures necessary for a new set of cognitive or physical functions. Take, for example, a recent (from an evolutionary perspective) phenomenon such as reading, which uses older, existing neural structures in the visual system, such as those for recognizing contrast and sharp corners.[8]

A study of zebra finches from 2011 also offers an example of neural reuse. This study shows that the areas of the brain developed for singing are also involved in the act of avoiding certain food. With the help of state-of-the-art genetic technologies, the researchers were able to demonstrate that the neural areas involved in producing song (the frontal premotor nuclei) overlap with certain networks or circuits that had evolved earlier during the search for food. Apparently, certain food-searching mechanisms developed that later turned out to be useful for the "song control system" (a system of brain areas and neural pathways responsible for song learning) of birds.[9]

In this sense, we can compare neural reuse with what the evolutionary biologist Stephen Jay Gould called "exaptation," the use of existing traits for a new function. We see this process in the case of feathers, for example. Feathers originally served to regulate heat but later, as the capacity for gliding flight developed, were selected as an effective way of physically propelling the bird forward. During the course of evolution, old biological mechanisms are reused in completely different roles or functions, mostly without effecting any change in the underlying genetic structure.[10]

Something similar may have happened in the long history of the development of musicality and language. In a more recent phase, existing structures involved in musicality may have been partially reused and specialized as our capacity for language evolved. This is a neuroscientific interpretation of Darwin's assumption that musicality precedes both music and language functions. However, thus far it remains no more than an assumption. The challenge is first to identify the constituent functions of our musical ability, and then to study the extent to which those functions are present in our processing of music and language.

According to Darwin, sexual selection played a vital role in the origin of language, based on the ability to imitate sounds, often called "vocal learning ability," a skill that both humans and songbirds possess. Where musicality was concerned, Darwin was thinking primarily of song: "in producing true musical cadences, that is in singing." As well as playing a role in courtship and territorial behavior, musicality was mainly intended to express emotions such as jealousy, love, and victory. This notion, which has since been developed further by evolutionary linguists, has been dubbed "musical protolanguage" in the recent literature.[11] For reasons mentioned earlier, however, I will continue to call it musicality or musical ability.

To assess Darwin's assumption scientifically, the central questions remain: What is musicality? What are its components? Which neural networks does it involve? Answering these questions was precisely the aim of my quest, which by this time was well under way.

Querétaro, January 19, 2012. A neurobiological conference on "subsecond timing," organized by Hugo Merchant, began early this morning. This term refers to the study of short rhythms on a timescale that is also relevant for music, a timescale lasting about one second or less. Think of the beat in music (on average, about a half second), or of "expressive timing," the timing that allows one to hear whether someone has been trained as a jazz or a classical musician. These are time variations that take place on a much smaller timescale (between 50 and 1,000 milliseconds) than that of the beat. Throughout the day, psychologists, neurophysiologists, and other experts discuss the process of perceiving time and the neurology of this process. Our presentation is scheduled for tomorrow. In it, we will reveal the first results of the listening experiment with Capi.

The last eight weeks had passed by slowly. I could think of little other than Capi, who was busy listening involuntarily to my rhythms on the other side of the ocean. Every day, I had wanted to send Hugo an e-mail asking how it was going, but there wouldn't have been much point, because the test measurements hadn't been completed yet. Every day, though, a new file appeared on our shared data server. With special computer scripts, we checked the log files in Amsterdam to see if the results were already pointing in a certain direction. It was like monitoring a lunar probe.

Hugo had conducted five extra sessions with Capi in the week before the conference in the hope of obtaining even clearer results. Between the lectures, I consult Hugo to see whether the conclusion we hope to present is still supported by the most recent data and analyses.

So far, there have been strong indications that Capi recognizes the beginning of the rhythm (marked by a brief silence), and also all the silences, but not—as was the case with the newborns—the silences occurring on the first count of the beat more than those at all other positions. Our provisional conclusion is therefore that we *can* measure an MMN (mismatch negativity) in reaction to unexpected tones, but that Capi's reaction to the "silent" and "loud" rests does *not* reveal any beat perception.

Our colleagues at the conference seem convinced by the data we present, but the criticism we were expecting is also voiced: "Yes, but $n=1$!" We had only tested one rhesus macaque! What if, by chance, we had been dealing with the Mathieu of rhesus macaques? Is Capi the exception or the rule?

After the conference, and after we have discussed our fellow researchers' comments in depth, Hugo decides to proceed with the listening experiment, this time with a second rhesus macaque. I am relieved, because this was not a given. For every repeated experiment, Hugo has to reconsider whether it is justified to use a new test animal. Though in exceptional cases the results of an experiment with one monkey *are* published in reputable journals (such as those of the previously mentioned chimpanzee Mizuki), for most neurobiological studies, it is more common to compare results obtained with two monkeys, and usually no more than two. Because of the strict guidelines and other understandable limitations, many researchers have to act strategically when using test animals in their experiments. Which research project has the greatest prospect of being reported in the leading journals? Which experiment entails risks but offers the greatest chance of a breakthrough? Or, perhaps, which studies are less innovative but may represent a significant step forward in the current research? And, last but not least, do the current findings justify using additional test animals?

Music and musicality are seldom the subject of articles appearing in the leading scientific journals. Though more and more fundamental research is being conducted on these topics, musicological research lags behind considerably, compared, for example, with linguistics. Many researchers still view music as a luxury, a cultural by-product of more fundamental phenomena such as language, speech, or cognition in general. It therefore means a lot to me that Hugo considers my research question important and believes in the design and the outcome of the tests. Hopefully the tests performed in his laboratory will bring me several steps closer to answering my questions regarding the biological foundations of musicality. And hopefully they will also provide the not-insignificant secondary result of having demonstrated that noninvasive methods like the EEG *can* be used for neurocognitive research.

The new rhesus macaque is named Yko. He has been taking part in different experiments in Hugo's laboratory for several years already. Last Saturday, after the visual experiment in which Yko normally participates, he was taken to the listening room so that he could start adjusting to a different daily rhythm. Next week he will begin with the dozing-listening sessions.

Hugo and his team will continue to take measurements in the weeks ahead. In two days' time, I will return to the Netherlands.

Amsterdam, April 4, 2012. "Something's not right," says Gábor firmly, as we sit opposite each other at a big table in my office. The table is scattered with printed tables and graphs. These are the measurement results for Yko and Capi that Gábor has analyzed diligently in different ways over the past few weeks. Contrary to our expectations, the statistical tests show that at certain pitches, the peaks in Capi's brain signal are unmistakably weaker than at most other pitches, a difference one would not have expected on the basis of the stimuli. After all, the rhythm remains the same, whether the tones are high or low. In the first two experiments, the signals of the two rhesus macaques appear to resemble each other, but surprisingly, this is not the case in the third and definitive experiment with the drum sounds.

Later that evening, I have a video conversation with Hugo and explain our concerns. We agree that the calculations are correct, so something else must be happening that accounts for the differences. Hugo proposes making audiograms (graphs from which the hearing sensitivity can be read) of the hearing of the two rhesus macaques. Is the hearing of both individuals intact?

A couple of weeks later, I find two urgent e-mails from Hugo in my inbox. It is immediately clear why: the audiograms have revealed that Capi is partially deaf. His hearing shows a small drop precisely in the frequency range where the drum rhythms are heard. This explains why the rhythms were clearly audible in the first two experiments but not in the last experiment. The upshot is that Capi's measurement results have become unusable for our purposes. All those weeks of listening and analyzing have been in vain. We should, of course, have tested the hearing of the rhesus macaques at the outset.

Though a bitter pill to swallow, the discovery also inspires confidence in the method and procedures that we are using. It plainly demonstrates the importance of "blind" analyses of the data and statistics as a way of ensuring that, as a researcher, one does not become too entrenched in one's own vision owing to well-intentioned enthusiasm. It also turns out that the method can indirectly test hearing.

Fortunately, Yko, the other rhesus macaque, has almost perfect hearing. We now know for certain that the measurement results from his listening sessions are reliable. But it also means we have to repeat Capi's listening experiment with yet another rhesus macaque.

4 Measuring the Beat

I had become curious to understand more about the significance of sound for rhesus macaques in their natural habitat. Although they are confronted with sounds on a daily basis in the laboratory, it struck me as important to examine the role of sound and musicality in their life in the wild.

Not all primate researchers agree, but it appears that, generally speaking, most Old World primates show little interest in sound, let alone music. Of all their senses, seeing and smelling have much more important functions. Numerous studies of rhesus macaques indicate that their limited repertoire of noises serves mainly to signal either a threatening or a submissive stance. The noises they make play a significant role in determining and maintaining hierarchy in the group. Stare straight into the eyes of a rhesus macaque, as I did with Capi, and it will instantly feel threatened. The animal will grimace, bare its teeth, and start growling. The emotions of rhesus macaques can be read easily from their faces (by humans and rhesus macaques, that is), and their vocalizations add little to this picture.[1]

Rhesus macaques are social creatures. They make friendly noises, such as lip smacking. This behavior is particularly important as it signals friendly intentions. A lot of lip smacking occurs when two rhesus macaques approach each other, as well as during grooming and in other social situations.[2]

In addition to growling and lip smacking, female rhesus macaques make another distinct noise, known as a "girney." It is a soft, nasal, melodic cry which they use to attract the attention of the offspring of other female rhesus macaques, probably to assure their mothers that they mean no harm. The girney is sometimes compared to the human behavior of talking to infants and small children with an exaggeratedly rhythmic and highly melodic intonation, known in the jargon as infant-directed speech (IDS).

All these noises, in both rhesus macaques and humans, may be remnants of what, in evolutionary terms, is an extremely old phenomenon, one we share with many other animal species—a precursor of language and music that allows us to exchange emotions and warn or reassure other members of our species. This interpretation can be viewed as further support for Darwin's assumption.

It is also possible, though, that girneys are merely an expression of excitement at seeing an unknown youngster, because, strangely enough, rhesus macaques do not use girneys when communicating with their own offspring. Girneys may simply be a way of staying on good terms with other mothers: "What a lovely child you have!" With their own offspring, mothers usually limit themselves to the lip-smacking sounds of friendly intentions. Together, these three types of speech sounds—growling, lip smacking, and girneys—represent an extremely small repertoire compared with that of other vocalizing animals.[3]

Acknowledging that sound is relatively insignificant for rhesus macaques makes any question about their possible musical preferences seem rather farfetched. Nonetheless, small-scale research has been conducted on the potential musical preferences of nonhuman primates. The best-known article, by researchers at Harvard University, has the revealing title "Nonhuman Primates Prefer Slow Tempos but Dislike Music Overall." If nonhuman primates are made to choose between different types of sound—marmosets and tamarins were the test subjects here—they prefer the slow rhythmic chatter of other members of their own species. They show no interest at all in music.[4]

The same result emerged from studies with other species of primates. Gorillas, for example, in the Buffalo Zoo in New York state became restless when piano compositions by Frédéric Chopin were played in their enclosures, whereas they were calmer than normal when exposed to ambient sounds such as those of the rain forest.[5]

Although scientists often cite the Harvard research article, it has also been heavily criticized. The cotton-top tamarins were not exposed to the right music. The experiment used only human and Western music, from pop music to Mozart, but also German folk songs. The tamarins avoided Mozart the most. This may be disappointing news for human music lovers, but, of course, we do not know what these monkeys really hear and appreciate.

The primatologist Frans de Waal was also bothered by the Harvard study's choice of music samples and the conclusions it drew based on them.

Together with his student Morgan Mingle, de Waal investigated how chimpanzees react to non-Western music. In his view, the findings of the Harvard study were not generally applicable because the experiments always used the same type of (Western) music. In a press release issued simultaneously with his research group's article, de Waal proposed that chimpanzees may dislike or be put off by Western music only.[6]

De Waal's research group alternately played African, Indian, and Japanese music near the chimpanzees' outdoor enclosures for twelve consecutive days. In one part of the enclosure, the chimpanzees were exposed to the music for about forty minutes a day, while at other locations the music was almost inaudible. By observing where the chimpanzees stayed while the music was playing, the researchers were able to draw conclusions about how pleasurable (or irritating) the chimpanzees found it. It turned out the chimpanzees were neither bothered by, nor attracted to, the African or the Indian music but they clearly avoided the locations where the Japanese music could be heard.[7]

The reader undoubtedly has some notion of what African, Indian, or Japanese music sounds like, but I doubt very much if the same is true for chimpanzees. As with the Harvard study, the research methods used by de Waal say nothing about the basis for the primates' preferences. Exactly what they listened to is still unclear.

Nevertheless, the Harvard researchers were adamant in their interpretation and conclusions. According to them, the primates avoided the music of the Japanese taiko drummers because, like so much Western music, it was regular and rhythmic. That regularity could be perceived as threatening because it evokes associations with the rhythmic pounding, such as chest beating, which chimpanzees themselves sometimes display when they wish to assert their dominance.

This is, however, just the researchers' interpretation, not one that is substantiated by the experiment itself. It is unclear which aspects of the music the chimpanzees found appealing, irritating, or tolerable. It could have been anything: melody, timbre, rhythm, timing, or the dynamic development of the music. For this reason, one cannot speak of musical preferences in chimpanzees. What was tested here was not so much a preference for the music of a particular culture as a sensitivity to a whole range of acoustic features. In this sense, the article's title, "Chimpanzees Prefer African and Indian Music over Silence," was an overstatement. The chimpanzees probably had no idea at all what African or Indian music did or could even sound like.

An important question therefore remains: when does what we humans consider to be music sound like "music to the ears" of other primates?

Monkey Keeps the Beat

In 2011, when I first visited Mexico, the notion that nonhuman primates have beat perception was far from proven. In fact, the few studies conducted up until then indicated quite the opposite. Hugo Merchant, for example, had already demonstrated in 2009 that a rhesus macaque could not be taught to move a joystick back and forth synchronously to the sound of a metronome. Rhesus macaques cannot anticipate the way humans can.

Other researchers, however, including me, remained convinced that nonhuman primates must have beat perception. Some researchers worked hard on experiments to demonstrate this theory. Others were less patient and presented their first impressions at conferences.

Patricia Gray of the University of North Carolina, Greensboro, is a case in point. She has a long-standing interest in music and biology and is convinced that nonhuman primates, particularly bonobos, experience a pleasure comparable to that of humans when they are beating a drum, and that they can clearly synchronize to the beat of music.

To demonstrate this idea, she had a special "bonobo-proof" drum built, one that was resistant to jumping, biting, and other similarly boisterous behavior. She then let the bonobos spontaneously drum along with human drummers sitting in an adjacent room. The heading of the press release issued later by Reuters read: "Bonobos, like Humans, Keep Time to Music." It was big news and received worldwide coverage. Nonhuman primates have beat perception too!

However, when I asked the authors for the original article because I was curious about the methodological details, to my surprise it turned out not to have been published. Despite repeated attempts, it had not been accepted by a scientific journal. Apparently it had not been found sufficiently convincing by colleagues.[8]

This made our listening experiment with rhesus macaques all the more urgent. After all, we still faced several outstanding questions: Can a nonhuman primate hear the regularity (the pulse) in a varying rhythm, as we had already shown adult humans and newborns to be capable of? Or would rhesus macaques listen to music as if it were ambient sound, with no attention to the regularity that humans appear to consider so important?

Thessaloniki, Greece, July 26, 2012. I am sitting in a stifling hotel room, preparing my lecture for tomorrow. As I put the finishing touches on the text, several swifts skim past my balcony at eye level. Their extreme maneuverability and piercing calls always give me a feeling of intense joy.

This week I am attending an international conference on the theme of music cognition. At six o'clock tomorrow evening, I will present the preliminary findings of our research in a special session on rhythm perception.

I think back on the conversations I have had with colleagues during the past week. One of them, Jessica Grahn, an American neuroscientist, also has plans to study rhythm perception in rhesus macaques and, in particular, the architecture and location of the neural networks involved. She, too, suspects that there is a fundamental difference between perceiving and processing regular and irregular rhythms. It may be that the network that recognizes regular, beat-based rhythms is missing in rhesus macaques.

Jessica drew my attention to a recently published article from Japan, concerning a study that replicated the behavioral experiments conducted earlier in Hugo's laboratory with rhesus macaques. The article reveals that two individuals from a related species of macaque (*Macaca fuscata*) can tap along to regular rhythms containing intervals of approximately one second but not much less. Again, however, they can only do this reactively and without anticipating.[9]

I add a reference to this study to the slide show that I am preparing, then reread my notes. Will other colleagues at the conference agree with our interpretation of the findings? Might we have touched on something that further substantiates a difference between rhythm perception and beat perception? This was a difference I had already discussed with Hugo in Leipzig in 2011, based on the observation that, in humans, different neural networks are involved in processing regular (beat-inducing) and irregular rhythms. Or was it too good to be true?

In the final slide, I summarize the results of our three experiments. The finding of the first experiment should convince everyone. It shows that we can measure an MMN in a rhesus macaque. It is a new finding and is demonstrated for the first time in our study. If it also helps to increase the popularity of this noninvasive method, that would be a wonderful secondary result.

The second experiment, with the unexpected rest, is also convincing. Although we still need to replicate the effect in at least one other rhesus macaque now that Capi's test results have turned out to be unusable, Yko's brain appears to register unexpected silences.

Together, the first two experiments bring us to the right starting point for the third experiment, in which we let Yko listen to a complex, varying rhythm (as we had done earlier with the newborns). In this experiment, we detect no difference in the brain signals in reaction to any silence. For Yko, all the silences appear to be equally unexpected. He notices all of them, unlike the newborns, who had only noticed a silence on the downbeat. This means that Yko is insensitive to the regularity that human listeners perceive in the rhythm, and the "loud rest" is as imperceptible to him as all the other "silent" rests.

Although the findings are based on a series of experiments involving only one rhesus macaque, the conclusion we will present tomorrow will be the same as the one we reached in Querétaro last January, namely, that rhesus macaques do *not* have beat perception.[10]

Amsterdam, November 6, 2012. This morning, rather mischievously, I send Hugo an e-mail with the sound of a champagne bottle being uncorked. Our article has been accepted! It will be published in a few weeks' time, under the title "Rhesus Monkeys (*Macaca mulatta*) Detect Rhythmic Groups in Music, but Not the Beat," precisely one year after my first visit to Mexico.[11] Since adding the results for Aji, the third rhesus macaque who by now had taken part in our listening experiments, we have a clear conclusion: rhesus macaques do *not* perceive regularity in a varying rhythm.

I am proud of the result, even if it is not what I had expected. After all, with Darwin as the great inspiration, I had anticipated that rhesus macaques *would* have beat perception. Our research also demonstrated that the "heartbeat as the source of beat perception" hypothesis was improbable. After all, all mammals (including rhesus macaques and humans) hear the heartbeat in the womb. It seems more likely that specific neural networks enable beat perception, and that these networks are weaker or absent in rhesus macaques.

This finding also seems congruent with the theory of speech evolution. Research had recently demonstrated that the reason humans *can* talk while nonhuman primates *cannot,* is not because nonhuman primates lack an anatomical adaptation (such as a larynx, a tongue, or lips) but rather because they lack the neural control mechanisms that make speech possible. Here, too, it is the brain that makes the difference.[12]

While all of this may appear to be a negative outcome—rhesus macaques cannot do something that humans can—the insights contributed greatly to the advancement of theories about the biological basis for language and music. The ongoing challenge was to combine methods and techniques from different disciplines to achieve new insights. When our ten-page article appeared in the scientific journal *PLOS One,* I also realized it would only be a temporary milestone in what would undoubtedly still be a long journey to trace beat perception in nonhuman animals. It was now a question of waiting for the first replications and the detailed nuances of the research findings.

Apart from possible nuances, our article made the case for using noninvasive electrophysiological techniques in neurocognitive research. With these kinds of techniques, it should also be possible to take measurements outside the laboratory, in situations more natural than those when animals are raised in captivity. Take, for example, the type of EEG headsets now commonly used when playing computer games. In other words, even if the results of replication studies do not, or only partially, remain unchallenged, then at least I will not have subjected Aji, Capi, and Yko to my rhythms in vain.

Gradual Evolution

In the following months, I work with Hugo to reinterpret the abundant literature appearing on the subject of beat perception. Although Darwin suspected that all animals with a nervous system would have beat perception, this turned out not to be true, at least not for all primates. Should beat perception therefore be seen as a capacity that had developed only recently and gradually in the evolution of primates?

Rereading and reinterpreting the recent literature culminated in the formulation of what we called the "gradual audiomotor evolution (GAE)" hypothesis. Admittedly, it is not the most inspired name, but we based our hypothesis on the existing neurobiological literature, which suggested that the neural networks that enable beat perception in humans are absent or less developed in rhesus macaques (figure 4.1). In humans, this network connects the auditory system (hearing) with the motor system, which controls the movements of our limbs and mouth, such as clapping, dancing,

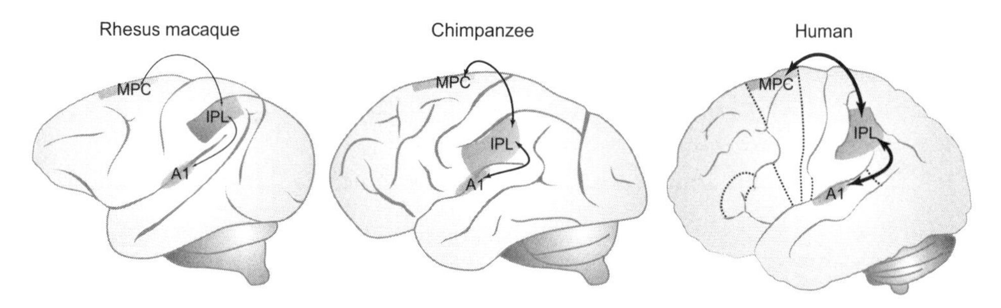

Figure 4.1
Diagram of the brain areas involved in beat perception in rhesus macaques, chimpanzees, and humans. The thickness of the line indicates the hypothetical strength of the connections between the most important areas (MPC: medial premotor cortex; IPL: inferior parietal cortex; A1: primary auditory cortex).

or singing. Even if you leave test subjects lying motionless in a functional magnetic resonance imaging (fMRI) scanner and let them listen to metrical and nonmetrical rhythms, activity is still visible in the motor cortex as a result of the metrical, beat-inducing rhythms. Clearly, an information exchange takes place between the auditory and motor systems.[13]

The absence of a strong connection between the auditory cortex and the motor cortex in most nonhuman primates may well be the reason why humans *do* and other nonhuman primates *do not* (or only to a lesser degree) have beat perception. We also proposed that this connection would likely be present in rudimentary form in chimpanzees, and therefore that chimpanzees would probably have beat perception in an embryonic form. If what we proposed was true, then we could date the origin of beat perception in primates to the time of the common ancestor of chimpanzees and humans, some five to ten million years ago. Of course, no study could be found to support this part of the hypothesis. It was therefore purely speculative.

Nevertheless, in my eyes, the GAE hypothesis still offered an attractive alternative to the "beat perception is uniquely human" hypothesis, which I thought was considerably less likely. After all, a striking number of the neural structures and networks that humans have at their disposal also exist in rhesus macaques and other nonhuman primates. In addition, rhesus macaques are no less skilled at "interval timing" than humans. Macaques can reproduce the interval between two clicks properly and accurately and also have no problem classifying a time span as either long or short.

What appears to be missing in rhesus macaques is the ability to perceive several intervals in succession. If a rhesus macaque is asked to perform a task in which not one but multiple intervals demand attention—as is necessary for recognizing a regular rhythm—it turns out to be difficult, if not impossible, for the macaque. A strong connection between the motor and auditory systems appears to be vital for beat-based timing or beat perception (see figure 4.1).[14]

Synchronizing

The musicality research agenda was gradually becoming more clearly defined, and interest in the cognitive and biological origins of musicality was growing, especially in the phenomenon of beat perception. More and more primatologists and other animal researchers were prepared to devote their precious research capacity to further probing this aspect of musicality. Their underlying motivation was to disprove the notion that beat perception was uniquely human. For that question, too, still remained: is beat perception uniquely human or not?

One example of researchers dedicated to investigating musicality is Yasuo Nagasaka and his colleagues at the Riken Brain Science Institute in Wako, Japan. They devised an experiment designed to assess the extent to which the ability to perceive "natural" or spontaneous regularity is present in the behavior of nonhuman primates. To this end, they placed two Japanese macaques (*Macaca fuscata*) in primate chairs on opposite sides of a table facing each other and taught them to alternately press the left and right buttons on the panel in front of them. If they did this regularly and at least ten times in succession, they were rewarded with a scattering of nuts or a piece of apple.[15]

The researchers were interested in *entrainment*, a natural phenomenon in which two oscillating systems match their periods and phases after a period of time. The Dutch physicist Christian Huygens (1629–1695), the presumed discoverer of this phenomenon, described it in 1665 after noticing that the motion of the pendulums of two wall clocks synchronized over time.

Entrainment can be demonstrated clearly with two metronomes, often used by musicians who are learning to play at the proper tempo. When two metronomes are set at the same tempo, they move back and forth at the same speed, or period. Rather than being synchronized at the beginning, they will be out of phase: the pendulum of one metronome, for example, will have swung only halfway across, while the pendulum of the other will already be swinging back. Surprisingly, if both metronomes are placed on a plank, itself resting on a surface that can move back and forth freely—for example, on two soft-drink cans lying on their side—after a short time the two pendulums will move into phase. For the remainder of the time, while rocking gently on the plank, they will continue to tick in perfect unison. This resembles the phenomenon of unconsciously walking to the beat of music that one hears on the street, regardless of whether one likes the music or not.

What is intriguing about such a complex and dynamic system is that it can be described clearly in mathematical terms. Calculations can be performed, precise predictions made, and specific observations assessed to see whether they meet the criteria of a similar system of coupled oscillators—in this case, the movements of two Japanese macaques.

The question here was: Could the macaques synchronize or "entrain" with the sound they heard in the same way that coupled oscillators do? And if so, did the macaques do so under the influence of what they saw, what they heard, or a combination of the two?

When the findings were analyzed, it appeared that the sound had virtually no impact on the synchronization. The Japanese macaques were therefore insensitive to the regularity of the sound. They did synchronize with the sound, however, when they were able to see the other macaque. In other words, the macaques were able to synchronize on the basis of what they saw, but not on the basis of what they heard. In this kind of experimental context, noise also turns out to be insignificant in terms of what Japanese macaques pay attention to.

In the months following the publication of this and other studies in 2013, the discussions that took place on the various scientific forums focused mainly on the question of whether the absence of beat perception in animals was categorical—some animals have it, while others do not—or whether there were gradations. If rhesus macaques, or macaques in general, did not have beat perception, then surely, I thought, chimpanzees and other anthropoids did. I went off in search of a research group interested in rhythm perception in chimpanzees. Two years later, I found it.

5 Ai and Ayumu

Inuyama, Japan, December 22, 2015. "Ai, Ai," resounds through the corridor, followed again, a short while later, by "A-i, A-i!" With her eyes cast upward, a Japanese researcher wearing light-blue cotton clothing, a surgical mask, and shiny white rain boots shuffles restlessly back and forth along the corridor. The ceiling comprises a steel grid that also serves as the floor of an ingenious network of corridors. This network connects the resident chimpanzees' indoor and outdoor accommodations with the different research areas. When an experiment begins, the chimpanzees are called in by name one by one. If they are in a playful mood or want a piece of fruit, they come quickly. At every junction in the overhead network of corridors, a small vertical trapdoor opens, allowing the chimpanzee to go right, left, or straight ahead. Yuko Hattori, my host today, controls the trapdoors remotely via switches in the corridors. Holding a small container filled with pieces of apple and other fruit, she walks in the direction of the laboratory, calling out as she does so. This is a signal for Ai, one of the oldest chimpanzees in the group, that all sorts of goodies are again on offer.

I had arrived at the Primate Research Institute (PRI) at 8:30 this morning. The institute is built on a hill on the outskirts of the city of Inuyama, not far from Kyoto (more centrally located on the island of Honshu). Yuko Hattori had just arrived by bicycle. She lives nearby and has worked at the institute for seven years. By this time, she has her own laboratory and specializes in the cognitive and social aspects of sound and rhythm in nonhuman primates. The other researchers associated with the PRI are particularly interested in attention, memory, language, numbers, and the visual system of these primates.

One of the PRI's best-known studies is an experiment in which chimpanzees have to remember the positions of the numbers 1 to 9, shown to them on a screen. The numbers change into square white blocks in a fraction of a second. The task is to remember the positions within that short period and subsequently to tap the numbers on the screen in the correct (ascending) order. A popular video of the experiment posted on the internet shows Ayumu, Ai's son, performing the task almost nonchalantly. In a fleeting glance, he registers the positions of the numbers in his working memory. I completed the experiment myself during the lunch break but lost track the moment more than five numbers were displayed, having flashed on the screen for less than half a second. This made no difference at all to Ayumu. He could instantly remember the entire image, sometimes with as many as nine numbers, and then, looking almost bored, tap the square blocks on the screen in the correct order.[1]

The experiment reminded me of how, earlier in the week, I had bought train tickets from a machine at a small train station where they displayed information only in Japanese. The layout was the same as for the machines in Tokyo, but they had also provided an English translation. I remembered the sequence from my Tokyo encounter and managed to touch the push buttons in the right order. Minutes later, I was holding the train tickets proudly in my hand. It immediately occurred to me that this task would pose no challenge at all for Ai and Ayumu. It was both a relativizing and a reassuring thought.

In addition to a large outdoor space, housing a fifty-foot-high, multilevel residential tower, the thirteen chimpanzees who live here have access to two adjacent buildings, allowing them to form new coalitions. Almost all the members of the group participate daily in an experiment in one or another of the research areas located on the site. As well as the thirteen chimpanzees, dozens of researchers work at the complex. They are emphatically referred to on the PRI website as *Homo sapiens*. The relationship between the researchers and their test subjects couldn't be clearer.

I am waiting in Yuko's research area. Shuffling noises can be heard in the overhead network of corridors connecting the laboratory with the outdoor area. Dressed in the special clothing that everyone, including me, is required to wear to prevent infection, Yuko enters the room, calling out all the while. With our surgical masks and other gear, we look like operating-room assistants. Yuko presses a button that slides open a trapdoor in the ceiling by means of a hydraulic pump. First one hairy leg drops down, then another. Gently swaying, Ai lands on the green-painted cement floor of the test area and walks calmly toward the Plexiglas window behind which I am sitting on a low stool. She presses her hairy gray chin against the Plexiglas and looks straight at me for a few seconds. Then she knocks on the pane once with her fist. Hey! I replicate her boxing gesture. We greet each other warmly.

Gray Fingers

Yuko Hattori is investigating whether chimpanzees have beat perception. But testing for beat perception in chimpanzees is no easy task. Measuring EEGs, as I did in humans and rhesus macaques, is not permitted at the PRI. The listening experiment performed earlier by Ueno with the chimpanzee Mizuki was an exception. Mizuki had been reared by human carers and had thus been conditioned to human presence. The chimpanzees at the PRI are left to their own devices as much as possible, with the researchers staying at a safe distance. An attached electrode brings all kinds of risks: it can be pulled off or seen as food or a new toy. For this reason, performing an EEG experiment on chimpanzees is out of the question.

While slicing an apple, Yuko tells me she expects to be able to demonstrate beat perception in chimpanzees. Like humans, chimpanzees live in relatively large groups, and beat perception may play an important social role. Synchronizing collectively with music appears to have an impact on what is known as "prosocial behavior." This can already be seen in young children: if they are allowed to move to the beat of the music during an experiment, they are more inclined to help the person running the experiment to pick up a pen that has "accidentally fallen" than a child who moves asynchronously or out of time with the music. Studies with adults also reveal that dancing or moving together to music enhances the group feeling and feelings of empathy. Deeply located brain structures such as the caudate nucleus appear to be involved. This structure, which humans share with many other animal species, forms part of the neurological "reward system" (a group of brain structures responsible for incentive). The reward system suggests an evolutionary function for beat perception: indirectly, by reinforcing feelings of empathy and solidarity, beat perception may have had an impact on the survival of the species.[2]

To investigate the extent of chimpanzees' sensitivity to regularity in music, Yuko developed an extraordinary behavioral experiment. She bought an electronic keyboard whose individual keys light up to help students who are learning to play the piano. Yuko demonstrates how the machine works by pressing the demo button. The instrument plays the chords of a tune while at the same time showing, via illuminated red keys, how to play the accompanying melody. This experimental design allows her to continue to build on what the chimpanzees have learned in other, primarily visually oriented experiments, involving, for example, touching an illuminated box on a screen. Before long, the chimpanzees learn to press the keys rhythmically on the electronic piano.

Ai stretches her long, gray fingers toward the piano keys. She has learned that she is expected to strike a key when it is illuminated. Another key lights up, which she also strikes. The moment she does, the first key lights up again. This pattern is repeated thirty times. In this way, Ai learns to alternately play a C, then a C an octave higher. Rather than Yuko or the electronic piano, Ai herself determines the rhythm and tempo at which she strikes the keys. At the end of each sequence, she is rewarded with a piece of apple, whether she has played evenly or not.

In an earlier study, Ai and two other chimpanzees had participated in a comparable experiment in which the regular ticks of a metronome were heard simultaneously on different speakers. Yuko was interested in the extent to which chimpanzees spontaneously began to tap the keys of the piano in time with the metronome. It was a way of demonstrating entrainment, comparable with Nagasaka's study described in chapter 4, but this time on the basis of what was heard, not seen.

The analyses of the timing of the piano playing revealed that none of the chimpanzees played spontaneously in time with the metronome. Only Ai did so regularly, and only at one specific tempo. However, that too may simply have been due to the frequency of her own arm (the tempo at which one comfortably moves one's arm back and forth). She didn't play synchronously with the metronome at any of the other speeds. For the time being, therefore, the experiment offered no clear evidence either against or in support of beat perception in chimpanzees.[3]

For today's experiment, Ai's son Ayumu is also present. While Ai strikes the piano keys earnestly and intently, Ayumu entertains himself in an adjoining area, hanging upside down from the ceiling and seemingly doing everything he can to distract his mother. He is unsuccessful. Ai pays attention to him only during her breaks, when she occasionally tugs on one of his arms or legs, but a moment later she is totally engrossed in the experiment again.

Ayumu's presence is part of Yuko's adapted experimental design. This time, it is not only about whether Ai plays the piano to the beat of the music—the task remains an unnatural one—but also about Ayumu's behavior in response to the music and the fact that his mother is totally focused on her piano playing. Ayumu begins to "pant-hoot" to some rhythms, softly at first, then gradually louder. It is a sign of excitement.

Through the loudspeakers, the same rhythms can now be heard that we had used in our earlier experiments with human newborns, but this time playing continually at different speeds. It is a clever trick on Yuko's part, intended to help her observe whether Ayumu will move in sync with his mother and, if so, at which tempo.

Yuko regularly makes notes about what is happening. She also films everything Ayumu does while his mother plays the piano. Later Yuko will analyze the videos to see if there is a relationship between Ayumu's movements, his mother's playing, and the tempo of the music. As yet, it is too early to draw conclusions; Yuko still has months of this type of experiment ahead of her. But from the stool where I am sitting, I have the impression that Ayumu *is* somehow affected by the rhythms and tempo of the music. While Ai continues playing, Ayumu sways back and forth in the adjoining area.[4]

Perhaps Yuko was right: having a comparable nervous system does not necessarily imply beat perception, as Darwin had assumed. It may be that it is precisely the social function of beat perception that has a decisive effect on its development. In that case, it should be possible to identify beat perception, and musicality in general, particularly in social animals. Rather than the result of a shared genetic basis, beat perception may be the result of an adaptation that, biologically speaking, can be achieved in very different ways. Such an adaptation may also have led to beat perception in animal species that are genetically more distantly related to humans (in a process known as convergent evolution).

I had discussed this idea earlier with several biologists and, around 2012, had quickly discovered that songbirds, especially zebra finches, were important animal models for learning more about musicality and convergent evolution. This realization turned out to be the beginning of a new, parallel adventure, because I had started my research on rhesus macaques at around the same time. However, in the case of my zebra finch research, in addition to the neurobiology tool kit, behavioral biology would now also play a key role.

6 Supernormal Stimulus

Leiden, October 31, 2012. The area where the Dutch behavioral biologist Carel ten Cate works is filled with a trumpet-like chattering—*trrrit, triiit, trrr, trrr*—fast, dynamic bursts of reverberating sounds. It is the chattering of the zebra finches that live in the two large aviaries here, with the males housed separately from the females. The incessant buzz reminds me of the excitement of teenagers at their first high school party. It is an infectious medley of flirting and dallying.

Around the corner is another area—filled with nesting boxes—that I am allowed to peer into quietly. A dozen or so zebra finches are brooding there undisturbed. The warmth and relative humidity are a telltale sign for the finches that the breeding season has arrived. Such conditions also mean the generations will succeed each other very quickly. A pair may produce several nests in a single year and sometimes even grandchildren. Because numerous generations can be studied during a short period, it is possible to trace the influence of the environment and the role of the genes on the singing behavior of these songbirds. The zebra finch (*Taeniopygia guttata*) is thus a much-used animal model for investigating whether various traits thought to be responsible specifically for language and speech in humans are also present in animals, and if they are, what that might say about the learning process (the ability to learn new sounds) and how speech might have evolved.[1] Using zebra finches for studying musicality was therefore a logical next step.

I have an appointment with Carel ten Cate at the Institute of Biology of Leiden University. The institute is housed in what, on first glance, is an unattractive university building from the 1970s, with rough concrete walls, recessed windows, and miles of suspended ceilings. But when I step out of the elevator on to one of the upper floors, I see that it has undergone ingenious architectural renovations. The restyled corridors are subtly illuminated with semiconcealed light fixtures and the light reflects off the shiny walls and white floors onto one side of the corridor. The walls are decorated with the same repeated image of a zebra finch perched on a stick and pecking at a red knob. It is a photograph of an experimental setup that can be found in numerous rooms here, areas where zebra finches have to work hard for their daily fare.

I have already spoken with Carel several times at length, mostly at conferences. It turned out we were both interested in rhythm and wanted to know to what extent humans share beat perception with animals. We had even written a research proposal together, although it didn't make it past the selection committee. The ideas were still too undeveloped, and virtually no results were available to suggest that the project would yield new insights. However, as is often the case, writing a research proposal makes one more convinced than ever about what one wants to investigate and why. We couldn't stop ourselves, and—overly curious as we were about beat perception in zebra finches—opted to start regardless.

At the time, there was much discussion in the scientific literature about the extent to which beat perception was uniquely human and, if it was not, to what extent the trait was present in other "sound imitators," that is, animals with vocal learning (vocal learning being defined as the ability to modify the acoustic or syntactical structure of sounds made by the vocal tract through imitation and improvisation). The American biologist and neuroscientist Ani Patel described the idea in 2006 in a short but influential article. He suspected that beat perception and the ability to move to the beat of music (synchronization) had something to do with vocal learning. He called his idea the "vocal learning and rhythmic synchronization" hypothesis.[2]

Patel's hypothesis proposed that only those animal species capable of imitating with their own voices and learning new sounds would be able to hear regularity in a varying rhythm and thus move to the beat of music. In animal species with vocal learning, the neural networks of the motor and auditory systems are connected in such a way that the species are sensitive to regularity in an acoustic signal (sound or music). In short, Patel's hypothesis suggested that vocal learning was a prerequisite for being able to hear regularity.

The "vocal learning as prerequisite for beat perception" hypothesis (hereafter the VL hypothesis) predicts that humans probably share beat perception with specific species of birds, such as parrots, budgerigars (common parakeets), and zebra finches, but not, for example, with horses, dogs, and nonhuman primates. The latter three groups do not have vocal learning, or, if they do, only to a very limited degree.

What is intriguing about the VL hypothesis is that it is diametrically opposed to Darwin's hypothesis, which suggested that the perception of rhythm was a trait that humans share with all moving animals. Bearing Darwin's hypothesis in mind, one would expect beat perception to have a long evolutionary history and to be a trait humans share with many other animals. The VL hypothesis suggests precisely the opposite.

Although most vertebrate animals can learn to recognize specific sounds, it is rare to find species in the animal kingdom with vocal learning. In fact, only eight animal groups are known to have it. The birds with vocal learning are songbirds, parrots, and hummingbirds. The mammals with vocal learning, in addition to humans (the only such species among the primates), are cetaceans (such as dolphins and killer whales), pinnipeds (seals and their close relatives), elephants, and bats (figure 6.1).[3]

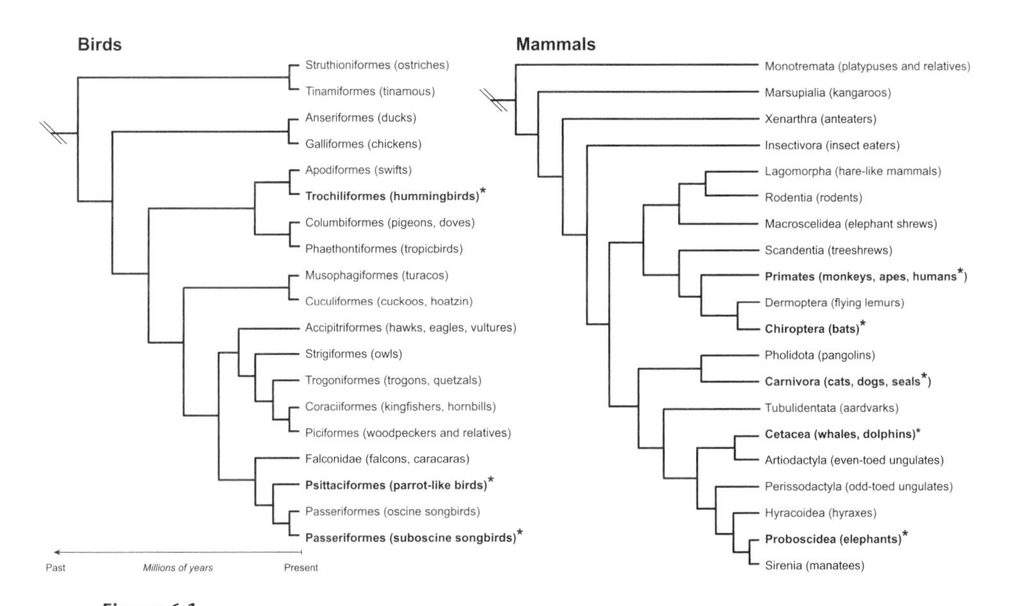

Figure 6.1
The eight animal groups (marked with an *) with vocal learning, and their evolutionary history. The birds with vocal learning are songbirds, parrots, and hummingbirds. The mammals with vocal learning, in addition to humans (the only such species among the primates), are cetaceans (such as dolphins and killer whales), pinnipeds (seals and their close relatives), elephants, and bats.

Most of these animals are only remotely related genetically. The common ancestor of humans and birds, for example, lived more than 320 million years ago. This is why biologists believe that vocal learning is a case of convergent evolution, in which a similar solution evolves independently for a similar problem, regardless of the biological genesis of the group of organisms. The physiological structure of the species may differ, but the function of the specific ability is the same (analogous).

Though controversial, scientifically speaking the VL hypothesis is a strong one because it is both measurable and testable. It is supported by research showing that specific animals with vocal learning also have beat perception. However, it can also be falsified: in fact, only one convincing counterexample is required—an animal without vocal learning that *can* synchronize to music—for the hypothesis to be relegated to the waste bin. From the scientific perspective, a falsifiable hypothesis is attractive. Put simply: not everything can be true.

The VL hypothesis further predicts that zebra finches should have beat perception because they too have vocal learning and learn their song from family members, usually their father. This is why I wanted to design a behavioral experiment with zebra finches together with Carel: to see if we could confirm the VL hypothesis. But before we could start designing a listening experiment, I thought it important to get a clearer picture of the methods Carel used in his laboratory.

In Leiden, later that afternoon. We walk to Carel's office, where several framed replicas of the famous herring gull heads belonging to one of his predecessors, the behavioral biologist and Nobel Prize winner Niko Tinbergen, are hanging on the wall. They are cardboard models of herring gull heads, each with a different-colored patch painted on the yellow beak. They are on display there as a tribute to Tinbergen's pioneering work and ingenuity. The replicas were carefully crafted a number of years ago by some of Carel's students when he repeated Tinbergen's herring gull experiment together with them.

Carel enthusiastically shows me several copies of Tinbergen's original notes, which he had inherited when he was appointed professor of behavioral biology at Leiden University. I immediately recognize the iconic diagram from the herring gull study, a stylized drawing featured in almost every behavioral biology textbook (figure 6.2).

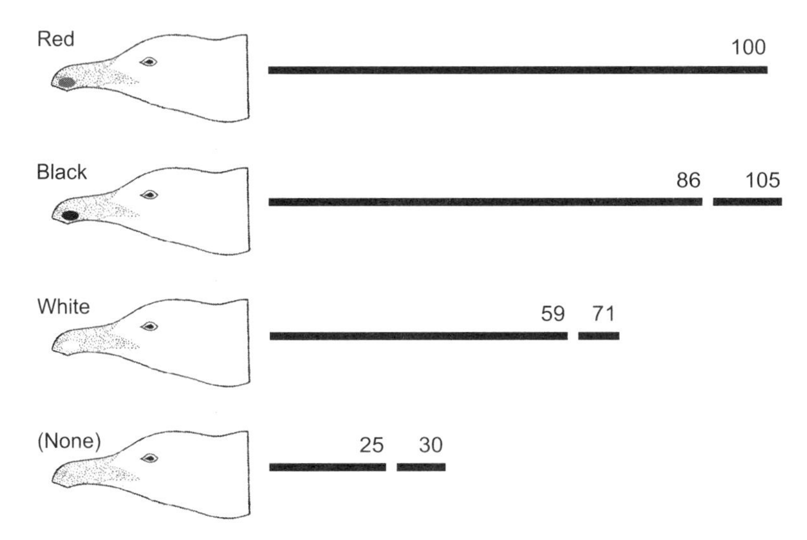

Figure 6.2
Pecking responses to the different herring gull models as described by Tinbergen and Perdeck (1950). The length of the black line corresponds to the number of pecking responses of the newborn chicks, expressed as a percentage in relation to the "naturalistic" model (model with red spot); the white breaks in the line indicate the values that were later adjusted downward.

Tinbergen was interested in animals' instinctive behavior at a time when scientists firmly believed that a lot, if not everything, could be learned from it. He had observed that animals come into the world with instincts that appear to already be adapted to the environment into which the animals are born. He wondered if the herring gull chicks' obsessive pecking at their parents' beaks, aimed at inducing them to transfer food, was instinctive, and what features of the beak elicited the begging behavior.

In a series of experiments, he presented a succession of different models of herring gull heads, with red, black, blue, or white spots painted on their beaks, to the chicks, then noted the number of pecking responses per half minute. This number served as an index, an experimentally established quantitative measure of the strength of the begging behavior: the more pecking responses, the higher the number. As simple as it was ingenious, the method made it possible to examine in an experimental setting what exactly elicited the begging response, something that would have been virtually impossible to observe in nature. In fact, the method allowed

Tinbergen to determine whether it was the color or the position of the spot, or possibly other features of the beak, that induced the pecking behavior.

He concluded that the begging response was primarily influenced by a combination of the spot's red color *and* the contrast between the colors of the spot and the beak. The big surprise lay in the contrast. Experiments had already shown that the color red caught the chicks' attention, but Tinbergen had observed a similar response to a white spot on a gray beak. As a result, in a follow-up experiment, he presented the chicks with a model in which both features—the color red and the contrast—were present: a bright-red pointed stick with three white stripes painted on it. The experiment revealed that this exaggerated (and also unnatural) combination of the most prominent features of the herring gull's head produced more pecking responses than any other model.[4]

Tinbergen termed this a "supernormal stimulus": the phenomenon whereby, when a normal stimulus is exaggerated, it becomes more attractive and thus elicits more responses than the natural version of that stimulus. This phenomenon occurs frequently in both animal and human behavior. With respect to humans, take the example of red lipstick, which tends to be considered more attractive and alluring than the natural color and shape of the lips. Or of all those things that are larger than normal, such as supersize fast food.

According to the evolutionary psychologist Steven Pinker, music could be a supernormal stimulus. He believes musicality is not the result of natural or sexual selection but rather a by-product of cortical systems developed for speech, among other things. These systems are supernormally stimulated by the tones from which music is constructed. In the same way as a herring gull chick finds an unnatural red-and-white-striped stick attractive, so humans find music beautiful. Pinker thus rather uncharitably characterized music as "auditory cheesecake." In doing so, he wanted to emphasize that musicality is not an adaptation but more likely a by-product of the—from an evolutionary perspective—infinitely more important "reward system," an old adaptation that still generates a positive response to the consumption of high-fat and sugar-rich food (hence Pinker's use of the word "cheesecake").

Despite all the criticism that Pinker received in response to his "music as by-product" hypothesis, there is something intriguing about his line of thought. Research has repeatedly demonstrated that music can appeal to the reward system, not only when we hear melodies that we think are beautiful, but also when we hear rhythms that we find enjoyable or appealing. Predictability and regularity give us pleasure: the brain produces the neurotransmitter dopamine when it hears the inevitable first beat (the downbeat) or that one expected crucial tone. In this sense, melody and rhythm together form a supernormal stimulus: an abstract stimulus that we call music, which can directly affect our biological system.

Darwin assumed that this "enjoyment" occurs in both humans and animals, and that, in addition to the perception of melody and rhythm (based on comparable neural networks), it is highly likely that enjoyment (elicited by the reward system) also plays a role in the origins of musicality.

Herring Gull Heads

A faded print lies on Carel's desk. It is the original of the classic diagram from Tinbergen's early publications. The document has troubled Carel for some time already. The original test results revealed that the black spot, not the red spot, elicited the most responses from the herring gull chicks. This was why Carel chose to repeat Tinbergen's classic behavioral experiments.

After his first experiment, Tinbergen realized he had made a methodological error: he discovered inconsistencies in the way the experiment had been performed. He had presented the chicks alternately with a red cardboard model and a model with another color, using the red model each time as a reference point for the attractiveness of the other colors. Red was therefore presented to the chicks much more often than the other colors. Tinbergen realized that this setup would make the "naturalistic" red model less attractive, because it didn't result in a food reward. This had to explain the lower score for the red spot compared with the score for the model with the black spot.

Based on a subsequent experiment, in which Tinbergen continually alternated the model with the red spot and a model with a black spot, he calculated a correction factor that would rectify the relatively low preference for the red model in relation to the black. He then adjusted the results in all his later publications using this factor (see the vertical breaks in figure 6.2). Tinbergen referred to these adjustments in his earliest publications but later omitted any reference to them. Even in his widely cited books *The Study of Instinct* and *The Herring Gull's World*, he no longer refers to the correction factor; he even states that all the colors were presented equally often.[5]

Although Tinbergen clearly had no fraudulent intentions, the fact remains that the experiments were never performed the way he described them in his books. Eager to uphold the honor of the esteemed behavioral biologist and encouraged in this endeavor by his students, Carel had little choice but to repeat the classic herring gull experiment in accordance with prevailing scientific standards.

In the summer of 2008, he spent seven days on the Dutch island of Schiermonnikoog in the North Sea, together with five students, and repeated the experiment. He performed two versions: the first was exactly as Tinbergen had performed it, with the red model continually being presented together with the other models; the second was as Tinbergen had wanted to perform it but never managed to, a version in which all the cardboard herring gull heads were presented equally often (the red one as often as the other colors).

In the end, the first experiment replicated the results exactly as they had been recorded in the original table: the black spot produced the most pecking responses. The correct version of the experiment, however, which Tinbergen had planned in detail but never performed, showed a preference for the red spot. Carel was surprised that the results matched each other so closely. He had not expected that performing the experiment differently would have such a major impact. Tinbergen had calculated accurately. The classic example of a controlled experiment in behavioral biology turned out to have the same results as are reported in all the school and university textbooks.[6]

One might think that Tinbergen had performed his experiments sloppily at the time. But in the 1940s and 1950s, his methods counted as highly sophisticated systematic research. Moreover, it is unfair to apply today's scientific standards retrospectively to earlier pioneers like Tinbergen.

The case also illustrates the importance of replication, one of the pillars of scientific research. Unfortunately, not much has changed since Tinbergen's day with respect to replication research. Science continues to focus primarily on promoting new insights and less on testing and upgrading existing knowledge. In that sense, the replication experiment of Carel ten Cate's research group represented a laudable exception.

Replication and Falsification

In a new discipline, like ethology, for which Tinbergen was a trailblazer, replication is not the first thing the researcher engages in. There are, after all, so many new things to discover. The same applies to my discipline of music cognition. As a researcher, one always has to guard against being prematurely convinced of a specific result. Our research on human newborns is a case in point: it produced excellent results and received extensive media coverage, but it has yet to be replicated. The design and results of that experiment have, however, served as the basis for numerous follow-up experiments, including the later listening experiments with the beat-deaf Mathieu and Marjorie and the rhesus macaques Aji and Yko. The question remains, though: were the methods and results reliable enough to be built on so quickly?

Glenn Schellenberg, a Canadian psychologist, didn't think so. He was asked by the American Psychological Association to write a commentary in response to our study on newborns. In that commentary, he said he wasn't convinced of the results and pointed to a possible flaw in the method. In his view, there was a "confound" (an alternative explanation) for the mismatch negativity on the downbeat: the possibility that the peak in the brain signal, the MMN, had been caused by the omitted sounds could not be ruled out. If there was such a thing as an "acoustic afterimage" and the omission of a hiatus that would create a different kind of silence from the omission of the sounds of a bass drum, this could lead to an alternative explanation of the results.

We were, of course, fully aware of this possible—though, in our opinion, unlikely—interpretation. But the critique kept haunting us. Bearing Tinbergen's and Carel ten Cate's replication studies in mind, to remove all doubt and preempt any possible alternative interpretations, we decided to repeat the experiment with adult musicians and nonmusicians, but this time under different control conditions. If one is going to go public with an eye-catching research result that might be refuted, it is better to do the refuting oneself.

In our new listening experiment, we chose to measure the brain signal elicited in response to an omitted beat (an omission) in three different places in a drum rhythm. In two of the resulting rhythms, a beat was omitted in a metrically strong position, a type of omission also known as a "syncope" or "loud rest" because it is quite noticeable—that is, the rhythm falters slightly. We also omitted a beat in a metrically weak position, resulting in a rhythm that, according to music theory, is not syncopated and contains a "silent rest." All three silences resulted from the omission of exactly the same sounds, which is how we hoped to avoid the earlier critique.

Fleur Bouwer, a fledgling doctoral student in our young research group, saw this as an ideal first research project and enthusiastically set to work. The first part of the research would attempt to replicate an earlier EEG study with adults, specifically with musicians and nonmusicians; the second part was new and would compare the MMNs in response to the three different types of rests. If the brain signals in response to the three omissions did not differ substantially from one another, we would no longer be able to maintain that the experiment could be used to measure beat perception.

Amsterdam, January 9, 2014. Today I received a short but enthusiastic e-mail from Fleur Bouwer. After a few minor textual adjustments, our study will be published in *PLOS One*. It is one of the newer scientific journals that claims to select articles more on the basis of the quality of the research and methods used than on the basis of the innovative ideas, a policy that allows replication studies, often not accepted by the more established journals, to still get published.[7]

In her first scientific article, Fleur showed that some of our earlier results could not be replicated. The difference in the response to a "loud rest" on the downbeat and to a similar "silent rest" halfway through the beat could not be replicated with a new and larger group of test subjects. This was contrary to what we had shown in a previous study.[8] Although earlier statistical analyses had suggested that we could be reasonably certain of replication if we used a large group of test subjects, this turned out not to be the case. Because of the statistical peculiarities of *p*-value procedures in empirical studies, this is a common error of judgment in such studies.[9]

Fortunately, the most important part of the experiment did have a favorable outcome. This time, with the test having allowed for Schellenberg's criticism, we observed a notable difference between the brain signal in response to the loud rest and the signal in response to the silent rest. In the end we found no difference in beat perception between the musicians and nonmusicians. This reconfirmed our earlier interpretation that musical expertise—regardless of how much music one makes or listens to—has no appreciable effect on the ability to hear regularity. In fact, *everyone* can hear regularity, even beat-deaf listeners like Mathieu and Marjorie (although they have no conscious access to that information).

More than anything, I am relieved, as it might well have been that the results showed no differences. Such an outcome would have undermined all our earlier work and cast doubt on our exciting results with the newborns, the beat-deaf Mathieu and Marjorie, and the rhesus macaques Aji and Yko. In hindsight, I find this shocking, but the thought that someone else might have falsified our research was even more distressing.

7 Snowball

Barcelona, Spain, November 16, 2007. The CosmoCaixa science museum resounds with the Backstreet Boys' 1997 global hit, "Everybody." A public symposium titled "Music and the Brain" is being held in this brand-new venue. A sulfur-crested cockatoo with the endearing name of Snowball can be seen on a large screen, dancing energetically and adeptly to the beat of "Everybody" while perched on the back of a chair in a dimly lit living room. The cockatoo's enthusiasm and elegant bows when he is complimented by his owner at the end of the video earn him a loud burst of applause from the audience in the packed room. It is astonishing to see how much excitement and emotion the video elicits; many people are teary-eyed. It seems we are easily moved when an animal can do something we thought only we were capable of.[1]

That evening, Isabelle Peretz, Robert Zatorre, Tecumseh Fitch, and I chat in a comfortable taxi van taking us to a restaurant in the city center. The discussion focuses on the YouTube clip that caused such a stir during the symposium. Tecumseh Fitch, an American cognitive biologist interested in the evolution of language and music, had recently been sent the video and had used it as the provocative end to his lecture. We are aware that the video may be a pivotal document. Were we witnessing the first video images of an animal with beat perception? Or was something not quite right? Had the clip been manipulated and the music only added later?

During the taxi ride, Tecumseh grows increasingly enthusiastic about the possible implications if the video proves to be authentic. It will support the VL hypothesis. But not only that. It might also mean that animals with vocal learning, such as hummingbirds, bats, walruses, and dolphins, have beat perception too. We must try to investigate it quickly. It may signify a revolution, a move in support of Darwin's assumption, as well as a rigorous challenge to the still widely held view that music and music perception are as yet exclusive to humans.

The other passengers continue to resist the idea. Robert Zatorre, a Canadian expert in the field of the neurobiological origins of speech and music, is mostly laconic: "I still don't see a walrus dancing to the beat of music."

"Has anyone ever tried it?" Tecumseh replies provocatively.

It is the video itself that had raised doubts in my mind. I wonder about the shadows I saw moving on the wall. Is somebody standing behind the camera and also moving to the music? Is the cockatoo not just imitating his owner?

We are clearly not going to see eye to eye, so we decide to make a bet: Is Snowball really listening, or is he simply mimicking his owner? The winner, we agree, will get a bottle of Marqués de Riscal Reserva, an exceptional local Catalan wine.

Why Birds Sing

To learn more about zebra finches—which, together with knowledge of rhesus macaques, is hopefully going to help me gain insight into the biological origins of musicality—I read many fascinating scientific studies of birds, particularly songbirds. One of my first questions is: Why does a songbird actually sing? And is what we consider to be song also *intended* as song?

Birds use sound for everything, from detecting enemies to foraging (think of owls) or identifying members of the same or a different bird species. To do any of these things, a bird must be able to distinguish between sounds (what *is* a member of the same species and what is *not*) and to locate where a sound comes from (where is the predator and where the mouse?).

Broadly speaking, birds' hearing is comparable to that of humans (figure 7.1). But there are also differences. Apart from the fact that birds, for example, do not have an auricle or ear flap—the auditory canal is mostly covered with feathers—all kinds of anatomical differences are present in the ear itself. The most interesting difference, though, is that birds' hearing fluctuates during the course of the year. In particular, birds living in a temperate climate experience seasonal changes, changes that appear to be caused by the length of the day and the effect that has on the birds' hormonal system.

The American biologist Jeff Lucas discovered that changes occur in birds' ability to both hear and process sound. He measured these two aspects

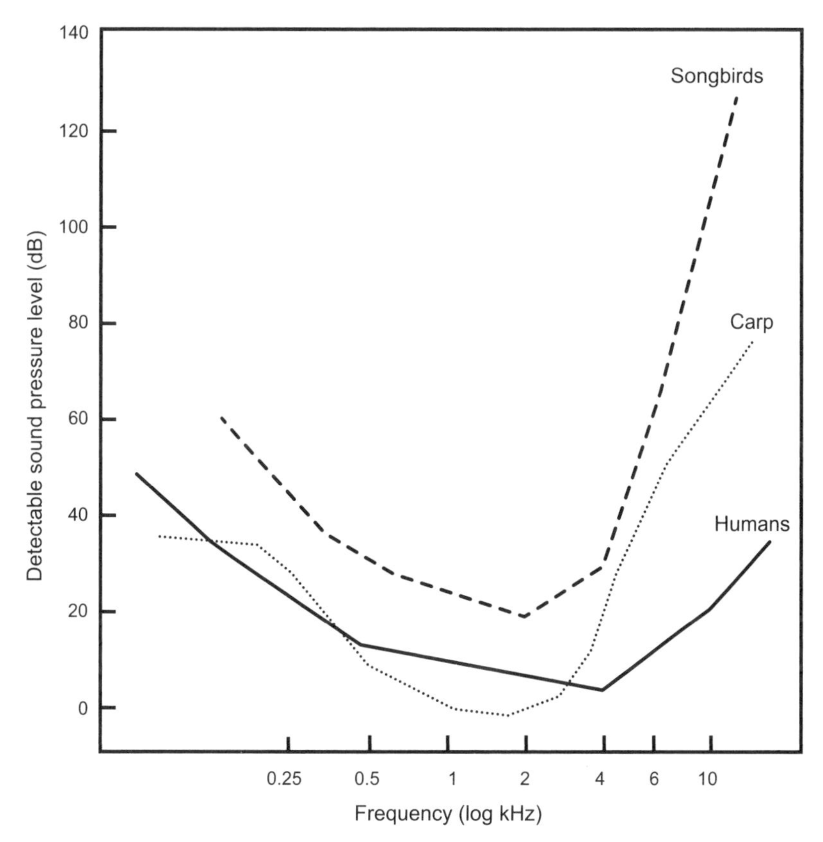

Figure 7.1

Audiogram showing average audibility curves for humans, songbirds, and carp. The lines indicate the hearing threshold: the sound pressure level of a barely audible tone as a function of the frequency. Songbirds hear best at around 2–3 kHz, and humans at a slightly higher level, at around 4 kHz. (Illustration after R. J. Dooling, "Audition: Can Birds Hear Everything They Sing?" in *Nature's Music: The Science of Birdsong*, ed. P. Marler and H. Slabbekoorn [London: Elsevier Academic Press, 2004], 206–225, fig. 7.2.)

using electrophysiological techniques (comparable to the invasive methods used in Hugo Merchant's laboratory) to study what aspects of sound are processed in bird brains and at what speed.

With the Carolina chickadee, the ability to process sound improves during the breeding season: they hear more details and nuances in sounds. In the case of the tufted titmouse, though processing does not improve, the birds are better equipped to detect sound: the nuances remain unchanged, but everything sounds louder and can thus be heard better. Finally, in the case of the white-breasted nuthatch, another American bird species, the sensitivity to sound increases within a highly specific frequency range. During the winter months, nuthatches hear more than normal around a specific frequency (2 kHz), probably because at about the same time, they are looking for a mate.[2]

Thus, in addition to playing a territorial role, sound figures prominently in partner choice. A study by the French ethologist Eric Vallet offers a good example of this phenomenon. He described what he termed "sexy syllables": the specific trills that make complex songs attractive to female canaries. If a male sings for a female during a period when she is fertile, she often responds by adopting a copulation-inviting crouching pose. This appears to be caused by a specific part of the song, namely, a series of rapidly alternating—about sixteen times per second—high- and low-frequency elements. Human ears perceive this as almost one sound, comparable to when someone sings a fast vibrato. Vallet's study, however, showed that the female canaries were able to hear the different elements of the trills: they could distinguish manipulated versions of the original song, in which several elements had been shortened or lengthened, from each other, and preferred the faster version, as reflected in the copulation-inviting crouching pose.[3]

Charles Darwin suggested that the origins of the precursors of language and music may have had something to do with sexual selection. He saw music as a "bunch of sexy feathers": developed not as an adaptation to survive but as an adaptation with which to attract potential partners.

The observation that songbirds can become excited by another bird's song may support this notion. In addition to the previously mentioned example of the canaries, female zebra finches also become animated when they hear the song of male zebra finches. They produce relatively more dopamine than normal in certain areas of the brain (more precisely, in the mesolimbic pathway). Dopamine is a neurotransmitter associated with the reward system, which is activated during eating and sex (see chap. 6). That birdsong can produce the same effect suggests that a biologically significant function is involved.

A similar mechanism has been found in humans. In listeners who said they got goose bumps when they listened to a favorite piece of music, an increased amount of dopamine was produced in their brain about ten seconds before the moment of listening. Memory plays an important part in this process. The brain looks forward, so to speak, to the listener's favorite moment—a wobbly note, an unexpected shift in the melody, a sob in the voice. And before that moment—roughly ten seconds beforehand—the body releases the same substance that the human reward system produces during eating or sex. It seems, therefore, that something as abstract as music also triggers the production of this substance.[4]

Let us return to birds, however. Songbirds can become excited when they hear a relatively complex song, for example, the "sexy syllables" that Vallet referred to. Some research even shows that male birds can become excited by their own song. According to the American neurobiologist Gregory Ball, singing stimulates the production of sex-related hormones such as testosterone, which then influence the form, structure, and chemical processes of the neural areas involved with song—not, therefore, the other way around.[5]

Ball believes that the chemical processes at work in a songbird—the hormonal system and the genetics underlying that system—are largely influenced by the singing itself. Rather than a sequence of "genetics-influences-brain-structure-influences-birdsong," he believes that a bidirectional interaction is much more likely, in which the bird's song has as much influence on the brain processes as the underlying genetics do. While only an observation, it sparks the imagination: birdsong as an abstract substance that can influence the plasticity of the brain.

Just as the hearing ability of birds changes under the influence of the seasons, which are so crucial to reproduction, so there are a growing number of indications that humans' hearing and experiencing of music can also change under the influence of hormones. Subtle differences can be measured in women's and men's hearing. In women, the hormone estrogen is apparently key to a periodically changing sensitivity in hearing. When the estrogen level is high, women perceive a man's voice as fuller. The effect is so subtle that most women are not even aware of it, yet it appears to play a role in partner choice and as such can be seen to support the notion that sexual selection is implicated in the evolution of musicality.[6]

In birds, the nature of that preference varies considerably. The Eurasian bittern (a type of heron) finds a low, deep "oong" attractive. The goldcrest (a member of the passerine family), on the other hand, finds the thin, high-pitched fluting of a potential partner appealing. Different qualities are therefore attractive for different species. What *is* shared, however, is the attraction of specific sounds.

This was emphasized again recently in an extraordinary study by Heinz Richner of the University of Bern in Switzerland. In his study he analyzed the relationships between the frequencies of the stereotypical two-toned song of the great tit. The relationship (ratio) between the frequencies of the two notes repeated by these birds in their song varies considerably. Expressed in terms of the harmonic series, it is usually close to a minor third and a major third (6:5 and 5:4, respectively), sometimes to a quarter (4:3) or a fifth (3:2), and only occasionally to a minor second (10:9) or a sixth (5:3). Richner demonstrated that the male great tits that sang closest to a rational number were found by the females to be the most attractive while simultaneously being perceived as threatening by other males. Interestingly, this is probably not because the birds pay attention to the pitch but because the song elements with low rational numbers resemble each other with respect to timbre or tone color more than those with more complex ratios.[7]

The observation that a bird's song or even a single call can be found appealing can again be seen to support Darwin's theory that musicality plays a key role in sexual selection. Some animal species, including humans, use it to impress their potential partners.

It is unclear whether sexual selection has also figured prominently in the evolution of music and musicality in humans. That music plays an important role during puberty—a time, after all, of massive development in terms of both identity and sexuality—is not necessarily proof of the notion that music results from sexual selection. And although famous musicians and especially pop stars have a reputation, because of their status, for having massive sex appeal and likely being much more sexually active than "ordinary" people, it turns out that on average they are more than twice as likely to die young. This is not something one can really call a reproductive advantage.[8]

Nevertheless, these are all indirect underpinnings of the sexual-selection hypothesis. One way of testing the hypothesis directly is to study humans as if they were birds, which is precisely what Tecumseh Fitch did.

In a listening experiment, Fitch asked women subjects to listen to different pieces of music, each with a different degree of complexity. The expectation was that the women would find the more complex music more attractive during ovulation than at other moments during their menstrual cycle. This hypothesis derived from the idea that songbirds select their mate based on the quality or complexity of their song (think of the "sexy syllables"). Although the women who participated in the experiment (forty in total) generally preferred the complex music, the results were not influenced by whether or not they were fertile. The experiment therefore did not provide support for the sexual-selection hypothesis.[9]

A follow-up study involving about 1,500 women made it possible to gather more precise data. When a distinction was made between choosing a partner for a short-term or a long-term relationship, the expected effect *was* observed when the choice was for a short-term relationship. In that case, the women who were at the peak of their fertility preferred the musicians, male or female, who made more complex music.[10]

The researchers interpreted this finding as offering the first empirical support for Darwin's theory that music also plays, or has played, a role in the evolutionary process of sexual selection in humans. But counterexamples exist too.

A major Swedish study examined the relationship between musical ability (in terms of both listening to and making music) and factors such as number of sexual partners, age of first intercourse, and number of offspring. The researchers investigated some 5,500 pairs of twins, for whom both the previously mentioned data and the DNA were available. A comparison of the DNA of the identical twins (who share almost 100 percent of their DNA) and the fraternal twins (who share 50 percent of their DNA) revealed that their listening skills and musical ability were only moderately influenced by genetic factors. This influence was equally significant for both sexes. The relationship with sexual activity was insignificant and sometimes even negatively correlated: generally speaking, musicians are not more sexually attractive or active and do not have more offspring than nonmusicians.[11]

In conclusion, the relationship between sexual selection and the attractiveness of musical behavior clearly differs in songbirds and humans. For the time being, we can only conclude that significant differences exist, and that the effect of sexual selection on musicality is not equally evident for all species.

Amsterdam, May 1, 2009. This morning I received an e-mail from Tecumseh Fitch, who has just been appointed professor at the University of Vienna. He refers to two scientific articles that appeared yesterday in the leading journal *Current Biology*. He writes, not without a degree of irony:

Dear Robert and Henkjan,

I remember placing a bet about the entrainment ability of Snowball the cockatoo, driving in a van during the music/brain conference in Barcelona, when I last saw you both.

I clearly remember the bet, and the reward for the winner—a bottle of Rioja (Reserva, no less)—but my memory of who it was who betted against Snowball's rhythmic abilities is a bit vague. I think it was one—or even both—of you. Please remind me if I'm wrong.

In any case, I'm less worried about the bet than the fact that, at least to my satisfaction, the issue has been resolved by two papers in yesterday's Current Biology *which pretty convincingly show an entrainment capacity in several parrot species. I found Patel's paper particularly clear and convincing, and hope you both agree.*

I look forward to toasting this advance in biomusicological knowledge at some future meeting!

Best, Tecumseh

Apparently Ani Patel had managed to make home videos of Snowball in collaboration with Irena Schulz, the cockatoo's owner and coauthor of one of the recently published articles referred to by Tecumseh. I found both papers on the internet. It seems this may indeed be a crucial breakthrough, based not on one systematic study but on two, of which the second also used a method considered unorthodox in biology.

Patel's article, the first, describes a simple listening experiment: Snowball was allowed to hear different versions of his favorite song by the Backstreet Boys, played a little faster each time with the help of a computer program. The program manipulated the tempo of the music while leaving the pitch and timbre intact. If Snowball really did hear regularity in the music, he should dance faster to the faster music. And that is exactly what he did, although only in about 15 percent of all the video recordings.[12]

The other article that appeared in the same issue of *Current Biology* was by Adena Schachner and her colleagues from Harvard University. They had analyzed about one thousand YouTube videos of nonhuman animals moving to music. Of all the animals the researchers found in those videos, only those with vocal learning were able to move synchronously to the music.[13]

The question remains, of course, whether this is a representative sample, but it is striking that both studies still supported the VL hypothesis. "Another one bites the dust," wrote Tecumseh Fitch triumphantly in his opinion piece introducing the two studies in *Current Biology*, with a reference to another of Snowball's favorite songs.[14]

The YouTube video of Snowball in action was probably one of the reasons for—if not the cause of—renewed global interest in the biological origins of musicality. In any case, for me, after the unexpected findings of our study on beat perception in newborns (also published in 2009), Snowball was the final nudge I needed to view biology as a necessary partner in helping us to understand what makes us musical.

The attention that *Current Biology* subsequently paid to the research—Snowball appeared like a pop star in different poses on the front cover—turned out to be a major stimulus for new biomusicological research. Since then, various research groups around the world have attempted to support or falsify the VL hypothesis.

Much discussion also took place about the methods used (they were relatively new and unorthodox) and the conclusions based on them. The informal design of the experiment with Snowball was criticized. For example, the researchers themselves had not been present during the experiment and had taken the owner at her word when she claimed not to have danced while the videos were being made. Hardly the most reliable way to preclude imitative behavior!

It later emerged that Snowball mostly danced when people were around, and preferably when his owner danced too. As is the case with humans and other social animals like chimpanzees, the social aspect is extremely important for cockatoos. This says something about the function of beat perception. Clearly the social component of moving to music cannot be overlooked. But this finding also makes it difficult to distinguish between moving back and forth rhythmically as imitative behavior or as a reliable measure of hearing regularity in music. Snowball's researchers wrote that the owner moved together with the cockatoo during the first two sessions, but not during the last two. As there was no statistical difference between the findings from the first and second halves of the experiment, the authors claimed that the owner's movements had no influence on Snowball's movements.

It was also strange to observe that a cockatoo that enjoyed dancing so much and moved to the *sound* of the music nearly the whole time, only moved to the *beat* of the music about 15 percent of the time. More comparisons with human dancing behavior were clearly needed. Perhaps humans danced asynchronously as often as nonhuman animals.

The authors of the groundbreaking article were, incidentally, fully aware of most of the methodological shortcomings and have since been actively engaged in seeking alternative ways of measuring beat perception in animals. The bet we had made in Barcelona, however, was still far from being won.

Metronome

One of the first more systematically performed experiments testing the VL hypothesis was undertaken by a group of Japanese biologists. Rather than analyzing video images, as Patel and Schachner did, Ai Hasegawa and his colleagues attempted to teach eight budgerigars to peck to the sound of a regularly ticking metronome. If the budgerigars pecked exactly to the sound of the tick six times in succession, they were rewarded with food, or, more precisely, with two kernels of grain. Budgerigars will spare no efforts when it comes to getting their grain.

Hasegawa used a common method known as "instrumental learning." Developed as a means of investigating the fundamental laws of learning processes, instrumental learning is distinct from evoked or reflexive behavior, the classic Pavlovian conditioning that we know so well. In instrumental learning, an initially neutral stimulus (such as the sound of a bell) is associated with a stimulus in the form of a reward (food). The bell announces the arrival of food, and thus, after a while, the response to that stimulus becomes the same as when food is offered.

Researchers can use instrumental learning to discover whether an animal has actually detected a stimulus, such as a photograph or a particular sound, and whether the animal can see or hear a similarity between a familiar and an unfamiliar stimulus. A bird is rewarded, for instance, when it pecks at a button after hearing a certain sound, whereas it should *not* respond (or be rewarded) when it hears other sounds. The method is primarily useful for establishing whether animals can distinguish between all kinds of stimuli (visual or auditory), how they do so, and if their responses resemble those of humans.

Hasegawa's study revealed that budgerigars can learn to tap synchronously to the tick of a metronome. They did so at different tempos, but most accurately at speeds similar to the rhythm of their own vocalizations. This resembles human behavior. Humans, too, are most accurate when perceiving and producing rhythms played at close to what is called their "preferred tempo." In humans, this is about 500 milliseconds on average, or about two beats per second (2 Hz or 120 bpm).

However, the information (included in the technical appendices to Hasegawa's 2011 article) on how the budgerigars pecked at the sensor revealed that it was difficult to establish whether the birds truly anticipated on the basis of counting, as humans do. When humans clap to the tick of a metronome (or to music), that movement begins *before* the tick so that it will be on time. If the precise timing of each clap is measured, one sees that the clapping begins, on average, dozens of milliseconds before the actual tick, in what is called "negative mean asynchrony" (NMA). In other words, humans *anticipate* the sound rather than *react* to it. With the budgerigars, the timing of the pecking was more or less simultaneous with the ticking of the metronome: they pecked just before and just after the tick equally often. This raises the question: were they responding to the sound in the way the rhesus macaques did, or were they anticipating the sound in the way humans do?

Computer simulations help to answer this question. They represent highly effective tools for identifying possible behavior that a pecking strategy can elicit and the statistical properties of that behavior. Do budgerigars peck as often just before, on, or after a tick as they would have had it been left to chance?

A computer program that simulates pecking on the basis of chance makes it possible to identify the likelihood that pecking will occur just before, on, or after the tick. Using simple statistics, it is then possible to establish whether the budgerigars' timing corresponds with, or indeed differs significantly from, the pecking strategy of the computer model. The latter turned out to be the case: the budgerigars did not just peck at random.

Another computer simulation used a pecking strategy where the pecking began as soon as possible after a tone was heard. Rather than responding to the regularity, the computer program responded to each tone individually, whenever it was heard. This "stimulus-response" model came closest to approximating the behavior of the budgerigars as a whole. The only difference was that, compared with the reflexively reacting model, the budgerigars occasionally responded slightly early.

By comparing various computer simulations with the budgerigars' pecking behavior, it was possible to eliminate the other alternative interpretations one by one, leaving just a single interpretation in the end: the birds anticipated the tick of the metronome and therefore apparently focused on the regularity of the tick.

However, one awkward detail in the experimental design remained: a little light continually flashed on and off to the tick of the metronome. This made it impossible to establish irrefutably whether the budgerigars perceived regularity through hearing or vision (imitation). Nevertheless, if one views the findings less strictly, they too offer support for the VL hypothesis.[15]

Leiden, October 31, 2012. After my conversation with Carel ten Cate about Tinbergen and his groundbreaking work in behavioral biology, we carefully return the tables and cardboard models to the cupboard and move on to discuss the design of our intended experiment. We want to test the VL hypothesis more systematically than existing studies have done. Ours is to become a pure listening experiment. In other words, there will be no analysis of movements in video recordings, as with Snowball, or allowance for alternative interpretations, such as recognizing regularity on the basis of what is seen rather than what is heard, as in Hasegawa's study. We are convinced that what an animal needs above all else to be able to move in time to the beat is the ability to perceive regularity. After all, the inability of human newborns and rhesus macaques to produce regular movements does not prove they cannot hear regularity. In our experiment, we therefore want to focus on the question of whether songbirds, specifically zebra finches, can distinguish between rhythm (such as a regularly ticking metronome or a clock) and an irregular, chance-based tapping (for example, the random spattering of an incipient rain shower) solely on the basis of what they hear.

For humans, this is a simple, almost trivial task. In fact, it is *too* easy for humans to hear the difference. Mindful of the experiment with the rhesus macaques, we also think it is wise to perform the experiment step by step, and particularly without setting our sights too high.

The first experiment will focus on the question of whether a zebra finch can distinguish at all between a regular and an irregular rhythm. Only in the second experiment will we test if the finches are able to "generalize," that is, to apply the rules they use for distinguishing between regular and irregular rhythms to unfamiliar rhythms. To test this possibility, we will use a design that has been applied more often in this laboratory: the "go/no-go paradigm," an advanced form of instrumental learning.

I am eager to know how the paradigm works. Carel suggests going to one of the rooms where a pilot experiment is being performed. We walk along a corridor lined with doors with blacked-out windows. Behind each door is a small, soundproof room. Each room is designed in more or less the same way: brightly lit with daylight lamps, longer than it is wide, with soundproofed walls that, for acoustic reasons, are not parallel to each other.

At the back of each room is a Skinner box: a spacious birdcage with numerous perches, a bottle of water, a piece of cuttlebone, and two sensors that operate a food dispenser. Above the cage hangs a small loudspeaker suspended by a cord from the ceiling. For the time being, it is quiet.

The room smells exactly as I remember our neighbor's canaries smelling. I took care of them once for several weeks when I was nine. Regardless of how pleasingly they sang, their company in no way compensated for my having to regularly clean their cage.

Here, though, things are different. The zebra finch shuffling up and down the perch has my full attention. I can tell that it is an adult male because of the orange patch on its cheek. It has a solid bright-red beak, a typical striped black-and-white feather pattern on its throat (hence its name), and subtle white spots on its chestnut-brown flanks.

Farther along the corridor, a barstool stands outside a room where an experiment is under way. Sitting on the stool, one can peer through the window right into the room. The piece of black plastic that normally obscures the window has been temporarily folded back. A zebra finch—a male named Bird 81—hops energetically back and forth from one perch to the other. Bird 81 started on a new listening experiment the day before yesterday. The data cabinets hanging on the back of the door indicate that it has already far exceeded the minimum learning criterion: the zebra finch learned surprisingly quickly which sounds do and do not result in food rewards.

The go/no-go procedure begins with the left sensor's little red lamp lighting up. It is a sign for the zebra finch that the session is about to begin. When the finch pecks at the sensor, a sound is emitted. Chance determines whether the sound is a go or a no-go sound. If it is a go sound, the finch must peck at the illuminated right sensor within six seconds. If it does so on time, the food hatch opens, and the finch can eat for a few seconds. If it is a no-go sound, the finch must restrain itself. Failure to do so will result in the light turning off for a few seconds, a signal to the finch that it has made the wrong decision. Unlike in a classic Pavlovian setup, in which animals learn that a relationship exists between sound and reward (or punishment), this setup validates making a distinction between categories of sounds.

When the birds first start on an experiment, they know nothing. They spend the first few days learning how the procedure works and are given extra food if they lose their way. A real zebra finch song is often used as the go sound, and an artificial pure tone as the no-go. Most of the birds quickly master the initial phase and, after a few tries, learn that pecking at a sensor only after hearing a go, not a no-go, sound results in food. The symmetry of reward and no-reward makes the experimental paradigm different from simpler forms of instrumental learning (as used by Hasegawa, for example). In the categorization phase—the classification of a stimulus into category A or B—the bird's consciously refraining from reacting to a no-go stimulus is as informative as its reacting to a go stimulus. Not reacting shows that the bird *is* capable of distinguishing between the two categories of sound, and makes the go/no-go paradigm a highly effective method for investigating all kinds of auditory, visual, and general cognitive functions in nonhuman animals.

We return to Carel's office to discuss the last formalities, such as the application for permission to conduct our experiment that we are required to submit to the university's experiments committee. The plan is to have four zebra finches take part in various related experiments aimed at assessing whether they can recognize regularity in a rhythm, as the VL hypothesis predicts. Because the experiment closely resembles previous experiments in terms of methodology, we do not expect the procedure to present any obstacles. I suddenly realize we have truly begun. My role is primarily to provide input on the stimuli that the birds will be exposed to. This time, they will include not "Everybody" or other global hits, but rather a systematically assembled set of rhythms from which we hope to be able to draw clear conclusions, in keeping with Tinbergen.

Amsterdam, April 20, 2012. Today Tecumseh Fitch is the guest speaker at an academic symposium on the similarities and differences between music and language. The lectures are being held in a packed hall in the heart of Amsterdam, and I will introduce Tecumseh as the guest speaker. It strikes me as the perfect opportunity to summarize our discussions of the past few years.

I show the by now all-too-familiar Snowball video again and explain why, as well as being delightful to watch, it was such an important discovery for our discipline. Apart from all the questions the video has given rise to, it remains an intriguing phenomenon: an animal that takes so much pleasure in the seemingly pointless action of moving to the beat of the music.

To emphasize the subject's topicality, I also describe Hasegawa and his colleagues' research findings on budgerigars that were published several months earlier. Hasegawa's study, too, lends itself well to a presentation. In addition to the clear experimental design, the videos are crucial here as well. They show a budgerigar tapping confidently and regularly to the tick of a metronome. I mention the problem of the flashing light briefly but notice that most of the students in the audience see this as splitting hairs. I had anticipated such a reaction and, as a result, had decided the week before that I would give Tecumseh the benefit of the doubt. If nothing else, this new research makes it highly likely that vocal learners, including parrot-like species, can perceive regularity. At the end of my introduction, I declare Tecumseh the winner of the bet we had made five years earlier, and present him with a bottle of Marqués de Riscal—"Reserva, no less."

Querétaro, January 26, 2012. It is dry and incredibly hot outside. I am sitting in a high-tech tour bus on my way from Querétaro to the Mexico City airport, heading home after my second visit to Hugo Merchant's laboratory.

On the bus, I use my tablet to search the internet for a recent broadcast of the Dutch science television program *Labyrint*, which ends with a segment in which viewers can ask questions of the guests. The episode I am looking for is about language and I am anxious to watch it, as Carel ten Cate was one of the guests. The previous fall, he and I had made our first plans to conduct research together on rhythm and beat perception in zebra finches. The television program focuses on various aspects of language and speech and on the zebra finch's role as an animal model for developing a better understanding of human language.

In the discussions, Carel explains that birdsong can be seen as a kind of chatter with which birds can impress potential partners and show where they come from. Not language, therefore, with words and clear semantics, but rather a kind of ... well, so far, a clear term is lacking.

As the interminable rocky fields and cacti fly past, I try to think of an appropriate term. Might birdsong be protomusic, musical prosody, vocal courtship, or, simply, the dialect of song?

8 The Dialect of Song

The songs of some songbirds consist of a single repeated element, while those of others, such as the nightingale, comprise a wide variety of sounds. Despite all the variation, however, the songs of many songbirds, including the zebra finch, are relatively constant. A song element is only occasionally omitted, repeated, or sung differently. Yet while the song is constant for any single zebra finch, it differs from one finch to another. The elements used are not the same, nor are they sung in the same order. With some songbirds, like the American white-crowned sparrow, this has led to "dialects," where certain elements and the order in which they are sung are determined by the dominant culture of the population.

In addition to song, songbirds and birds not deemed to be songbirds often have a repertoire of short sounds referred to as "calls." A call serves a more social function, such as announcing the arrival of food or inviting other birds to share it, warning about a predator, or maintaining social contacts within the group. In this sense, a call is (linguistically) more meaningful than a song.[1]

The British bird researcher Peter Marler, a pioneer in his field, demonstrated that thrushes and titmice make a high "*sip*" sound when they spot a hawk or other bird they consider to be dangerous flying overhead. The sound's high frequency ensures that the larger birds do not hear it clearly—they often hear high tones poorly—and a gradual swelling or diminishing of the sound's volume makes it difficult to locate. Think of the sound of a breaking branch versus that of a gust of wind rustling through the leaves. In the first case, one immediately hears where the sound comes from; in the second, identifying the source is not so easy. This is how birds minimize the attention they draw to themselves while at the same time warning other members of their species. Marler showed that a large number of bird species have this kind of meaningful call. It appears that both calls and songs are important components of a bird's life.[2]

Although terms like "dialects" and "sexy syllables" are problematic because they emphasize the linguistic aspect of birdsong, the question remains as to whether the order of the "syllables" matters and whether it influences the meaning of the song, as it does in human language. It turns out that variation within the elements does *not* result in a different function or meaning of a song, whereas in human language, the same elements (words) used in a different order take on a different—sometimes even the opposite—meaning (compare, for example, "man bites dog" with "dog bites man"). To date, no conclusive evidence has emerged to suggest that animals can change the meaning of their vocalizations by altering the order of the elements. I would not be surprised, though, if the components of such a trait turn out, in the end, to have been shared with other animals.[3] For the time being, however, it is wiser not to refer to the zebra finch's song as language but rather to emphasize its musical characteristics.[4]

That said, there are many reasons for comparing the twittering of zebra finches with the human capacity for language. Like humans, young zebra finches have a sensitive period during which they learn to imitate parts of their parents' song. Before they can sing an entire song, they begin with disjointed scratchings. Once they have the essence of the song under control, it becomes relatively long and structurally quite complex.[5]

Human language, on the other hand, is more than just word order and meaning. Language also has its own tone. Qualities like intonation and stress can make the same words interrogative, assertive, or ironic, for example. Linguists and phoneticists refer to these acoustic aspects of speech as "prosody" so as to emphasize their linguistic function. Yet there are equally good, if not better, reasons for referring to this sensitivity to patterns of intonation (melody), stress (dynamics), and rhythm as "musical prosody" or musicality. It is no mere coincidence that these patterns also form the building blocks of music.

In human development, this musicality is already active around three months before birth. Not only can unborn babies recognize their mother's voice and distinguish it from other voices, they can also remember melodies and, after birth, distinguish them from other melodies that they have not heard before. Both the perception and memory of melody are already functional during pregnancy. Unborn babies appear to listen mostly to the sounds as a whole, with special attention to the intonation contours, rhythmic patterns, and dynamic development of the sound. Only much later, when the infants are about six months old, does this musical prosody begin to play a role in what could be called the beginning of language, such as the recognition of word boundaries. During this phase of development, small tone curves, stress, and specific rhythms help infants to learn their mother tongue.[6]

From the perspective of developmental psychology, musicality thus precedes language. Infants' linguistic development benefits from the music-related sensory system for melody, rhythm, and dynamics, which has already been active for many months by the time they start to use it to learn language-specific categories such as phonemes, syllables, and words. All these findings point to the existence of a preverbal stage that predates both music and language.[7]

These observations are diametrically opposed to the idea that music is a by-product of language, a theory Steven Pinker proposed in his book *How the Mind Works*.[8] It appears more likely to be the other way around, namely, that musicality precedes both music and language, as Darwin assumed.

Leiden, March 1, 2013. Since I have been visiting the zebra finch laboratory in Leiden on a regular basis, I have become increasingly interested in birds. Is that a vulture or a buzzard I see on the pole over there? Is that a chiffchaff or a titmouse I hear in the bushes? A friend had given me a pair of binoculars for my birthday; clearly the people close to me had registered my changing interests some time ago already.

The aim of today's meeting with Carel is to decide which rhythms we will use in the planned go/no-go experiment. We need to compile two sets of stimuli, one with regular rhythms, the other with irregular rhythms. We listen over the small loudspeakers of a laptop to the different rhythms we had developed over the past few weeks on the basis of a growing list of criteria. Given all the experience I now have with rhythm experiments involving adults, newborns, and rhesus macaques, the pitfalls to be avoided are clear. Carel also has a list of dos and don'ts on hand and draws on his experience with earlier go/no-go experiments with zebra finches.

First, we choose only rhythms that comprise a short, percussive sound, so as to ensure that the rhythm will be clearly audible for the average zebra finch. In addition, we select sounds that will not have an obvious meaning in the bird's song repertoire. In other words, only the temporal structure—the rhythm—will change: the *what* will be wholly determined by the *when*.

To eliminate the possibility that the zebra finches will solve the problem by focusing on an aspect of rhythm other than the one we are interested in—the recognition or nonrecognition of regularity—we also know it is vital to ensure that the two sets of rhythms are as similar as possible and differ only in terms of regularity.

Our selection on this basis results in two sets of rhythms both of which have an equal number of beats and are equally long. Because previous experiments had revealed that zebra finches focus mainly on the first element or time interval of a sequence, in our further selection we take care to ensure that this interval is the same in both sets. All the rhythms differ from each other only at the third time interval.

Not surprisingly, it is primarily the irregular rhythms that are the most difficult to select. As it is virtually impossible for humans *not* to hear regularity and structure in an irregular rhythm, even if mathematically speaking they are not present, many alternatives had to be dropped before we could finally agree on a usable collection of rhythms.

Despite our efforts, we still wonder whether it is not far too easy to hear the differences. For us here at the table, it certainly is. Imagine, though, that you are a hungry bird: you hear two rhythms and don't know what to focus on to give the correct answer, an answer that will ultimately bring you food. Is it the irregularity of the rhythm, the grouping of the clicks, the total duration of the sound, or only the tempo that counts? Or is it the length of one of the time intervals?

The zebra finches will receive no instructions beforehand, the same as in experiments with humans. They will have to figure out for themselves what to focus on, and will only be helped by a reward or a disincentive telling them whether or not they are on the right track. Because of this setup, most of the zebra finches will take several days to master their role in a specific part of the experiment. For us humans, the task was perfectly clear: Does the rhythm sound like the regular tick of a metronome or not? But will the task be as obvious to the zebra finches as it was to us?

Timbre

Birds generally appear to perceive both the timing and frequency of sound more accurately than humans. It takes humans around ten milliseconds to hear whether two sounds are made simultaneously or in quick succession, whereas it takes songbirds around one or two milliseconds. The songbirds therefore make the distinction ten times faster.[9]

Songbirds also have superior hearing when it comes to timbre and frequency. For humans, timbre is usually secondary. It is like "filling in the colors"—not essential for recognizing the forms. Regardless of whether a melody is played on a piano or a violin, the melody sounds the same to humans. They recognize the different instruments, but the melody is crucial. The similarities (the pitches), not the differences (the timbres), are most striking to the human ear. For songbirds, however, even small differences in timbre matter, so much so that the birds would probably not even recognize a song as being the same, even if the melody was identical.

Admittedly, timbre is a difficult concept. It is generally defined in terms of what it is not: that aspect of sound that allows two sounds with the same pitch and volume to be experienced differently. In other words, when the same tone is played at the same volume on both a piano and a violin, the difference in sound is called "timbre."

At a very young age, humans can already recognize two sounds with a different timbre but the same pitch as having the same tone. Even newborns can distinguish between timbre and pitch. This ability indicates which aspects of sound are essential to them and which are less important.[10]

To humans, a harmony or a chord (such as a major triad) also sounds the same whether played on a piano or a guitar. Even without knowing whether it is a triad or that it is called an E-major chord, it still sounds identical. Humans appear not to distinguish chords on the basis of timbre. They disassociate from the timbre and concentrate on the pitches and the relationships between the pitches that comprise the triad.

If a similar test is performed on songbirds, they will *not* do this. The moment the timbre changes—and it only has to be a minute detail in the acoustic spectrum of that particular sound—the birds notice it immediately and are inclined to classify it as another sound, another song, or, in this case, another chord.[11]

Birds, primarily songbirds, use the vast richness of sound to gather information about other members of their species: Who is he? Where does he come from? Is something the matter? And, more fundamentally, is he attractive?

Songbirds focus on different aspects of sound than humans do. Whereas humans listen readily and very early on in their lives to melody and rhythm (rather than timbre), birds seem to concentrate more on the detail of each sound. In this sense, songbirds hear more than humans do, but they also hear differently in a qualitative sense. With songbirds, timbre and pitch appear to be closely intertwined.

As far as listening is concerned, one could say that humans are "supernormally stimulated" by the pitch of a sound (think of the red-and-white-striped stick in Tinbergen's herring gull study). They pay more attention to the discrete tones of a melody than to the continuum of the sound itself.

Another major difference between songbirds and humans is that humans can easily recognize a transposed melody as the same melody. They focus more on the melodic contours and intervals between the pitches than on the exact frequencies of each individual tone. For songbirds, the same melody sung at a slightly higher or lower pitch is an entirely new melody.

Songbirds can, however, recognize and even learn human melodies, as a group of dedicated musicologists demonstrated several years ago. Their study revealed that bullfinches can learn entire (sometimes even 45-note-long) German folk songs if the songs are whistled to them over a period of weeks. This finding illustrates the flexibility of songbirds' vocal learning abilities.[12]

But songbirds appear to focus on other aspects than the relationships between the frequencies, raising yet another interesting question: what do songbirds actually listen to when they hear that two melodies differ?

9 Perfect Pitch

In 2011, a video appeared on the internet with the viewer's comment "Look what this dog can do. He's more musical than I am!" The video showed a golden retriever looking intently at his owner sitting opposite him. When she produced a tone on a ceramic flute, the dog responded by striking a key on the large keyboard located on the ground in front of him. Regardless of the tone she produced, the dog repeated exactly the same tone on the piano without a moment's hesitation. "He has perfect pitch!" another viewer quickly noted. And it is true. Most animals have perfect pitch, in the sense that they remember and recognize sounds on the basis of absolute frequency (the vibrational frequency), not on the basis of melodic progression or interval structure, as humans do. This is why we think perfect pitch is impressive.[1]

If you ask musicians for an example of what they consider to be an exceptional musical skill, many will first say perfect pitch. Someone who has perfect pitch can name any randomly struck piano key without having seen which key was played. This is particularly useful for conservatory students because it means they have less difficulty with musical dictation—writing down in musical notation what the teacher plays or sings.

While I believe the golden retriever does recognize the flute tones by their frequency, I also suspect he would not be able to perform the task quite so perfectly if he and his owner were separated by a curtain. The eyes of the golden retriever and his owner are not visible in the video, but it wouldn't surprise me if the dog mostly follows her gaze to see which key he is expected to play. In that sense, he reminds me of the famous German horse Clever Hans (from the early twentieth century), who appeared to be able to do arithmetic and answer questions about numbers by tapping on the ground with his hoof a specific number of times. In the end, it turned out that Hans could not do sums but was extremely adept at reading the nonverbal behavior of his questioner, who of course knew the answer.

However, even if the golden retriever does perform the task primarily on the basis of perfect pitch, it is important to realize that he is doing more than just recognizing a tone by its frequency. In this respect, the term "perfect pitch" is too limited. As well as hearing, remembering, and recognizing a tone, the dog must also classify it (is it a C or a C-sharp?), decide which key it corresponds to, and subsequently strike that key on the keyboard in front of him.

In other words, perfect pitch is less an aural skill, as the term suggests, and more a cognitive ability. It has at least two dimensions: being able to remember a frequency and give it a name, then being able to classify that named frequency or pitch—whether it is produced by a piano, a violin, a voice, or a flute—as belonging to the same category of pitch and subsequently name it.

The first is a common skill that is easy to test. Imagine a familiar song such as "Stayin' Alive" by the Bee Gees and then sing it. Chances are the pitch will match that of the original perfectly. Like many other animals, humans are good at remembering the pitch of pop songs or familiar TV tunes. But hearing a single tone and then knowing whether it is a C or a C-sharp is a special skill possessed by fewer than one in ten thousand people.[2]

We have good reason to believe that perfect pitch results from a genetically determined predisposition. Studies show that certain chromosomes (such as chromosome 8q24.21) determine in part whether that skill is present or not. Neuroscientific studies also reveal anatomical differences, mostly in the temporal lobe and several cortical areas, between people with and without perfect pitch.

Not only does perfect pitch appear to have a genetic component and therefore to be hereditary, its development is also likely the result of both exposure to music at a young age and intensive musical training. Perfect pitch is much more common in Japan, for example, than elsewhere. Sometimes as many as 70 percent of Japanese conservatory students have perfect pitch, perhaps because music occupies an important place in young children's education in Japan.

As far as we know, though, perfect pitch has little to do with musicality. Generally speaking, people who have perfect pitch are no more musical than those who do not have it. In fact, the vast majority of professional musicians in the West do not have perfect pitch.[3]

Relative Pitch

At the beginning of the last century, Ivan Pavlov already discovered that dogs could remember a single tone and associate it with food, for example. It is also known that wolves and rats recognize members of their own species by the perfect pitch of their call and, therefore, that they can differentiate tones. Studies suggest that the same applies to starlings and rhesus macaques.[4]

A much more musical skill, though, is "relative pitch": recognizing a melody regardless of the exact pitch at which it is heard or sung. Most people listen not to a melody's individual tones and their frequencies but to the melody as a whole. Whether you hear "Mary Had a Little Lamb" sung at a higher or lower pitch, you still recognize the song. It is even possible you may hear a tune on the loudspeakers in a noisy café and still be able to recognize it instantly.

But who was the singer? You rack your brains, making associations in the hope of remembering the singer's name or the song title. When that doesn't work, you turn your smartphone toward the loudspeaker. The software gives you all the information you need within seconds. You now know exactly which song was played, who sang it, and which album it was from, a feat we find quite extraordinary.

To make this possible, the software producers have systematically analyzed and efficiently stored most of the commercially available recordings. A unique description of each song, an acoustic fingerprint that says something about the specific acoustic qualities of each piece of music, is stored in a huge archive. The computer program subsequently compares the fingerprint of the piece of music recorded on the smartphone with the one in the archive, then quickly and efficiently identifies the recording. While a piece of cake for computers, this task is virtually impossible for humans.[5]

However, if you hold your smartphone close to someone singing the same song, the software will respond by saying it has no idea what is being sung. Or it will make a wild guess. The version of the piece being sung is not included in the database of analyzed music, so the software cannot find the fingerprint. By contrast, humans placed in the same situation will recognize the song instantly, and the song may even resonate in their minds for days to come.

A computer would be surprised, so to speak, to learn that we need only half a song to identify who is singing it or what is being sung, regardless of whether it is sung at a higher or lower pitch, slower or faster, in tune or out of tune. For humans, part of the pleasure of listening to music derives from hearing connections and relationships (both melodic and harmonic) between the tones.

For a long time, scientists also believed that songbirds recognize and remember melodies based on the pitch or fundamental frequency, a skill that is a form of perfect pitch. The American bird researcher Stewart Hulse reached this conclusion some forty years ago after performing a series of listening experiments with European starlings. At the time, he showed that the starlings could discriminate between ascending and descending tone sequences, but not if the sequences were played at a slightly higher or lower pitch. Hulse concluded that the birds focused on the absolute frequencies. European starlings, like many mammalian species, turned out to have perfect pitch rather than relative pitch.[6]

Relative pitch, or the ability to recognize transposed melodies, has been well researched in humans. Neuroscientific studies reveal that relative pitch uses a complex network of different neural mechanisms, including interactions between the auditory and parietal cortices. This network appears to be lacking in songbirds. In researching the biological origins of human musicality, the absence of this neural network in songbirds makes the question of whether humans share relative pitch with other animal species all the more fascinating.[7]

As far as we know, most animal species do not have relative pitch. Humans appear to be the exception. One might wonder, though, whether relative pitch should be limited to pitch alone. Might sound have other aspects in which not the absolute physical characteristics but rather the relationships between them contribute to musicality?

In 2016, researchers at the University of California, San Diego, made an important contribution to providing a possible answer to this question. They exposed starlings to different melodies in which both the timbre and pitch had been manipulated. The stimuli consisted of what one might call sound-color melodies, tone sequences in which each tone has a different timbre. A series of go/no-go experiments studied the acoustic aspects of the melodies that were used by the birds to classify new, previously unheard melodies as either a go or a no-go stimulus.[8]

Surprisingly, the researchers discovered that the starlings did not use pitch to distinguish between a go and a no-go stimulus, as had previously been thought, but rather timbre and changes in timbre (spectral contour). The birds responded to a specific song even when it had been manipulated and all the pitch information removed using "noise vocoding" techniques. The resulting melody resembles a noisy sequence of sounds, a sound-color melody in which the sounds change from one note to the next but have no perceptible pitch. Only when little information remains, as with the stimuli in Hulse's European starling experiment (the stimuli consisted of pure tones, tones without any spectral information), do songbirds pay any attention to pitch.

Zebra finches have also been shown to focus mainly on the spectral contours, particularly when they need to distinguish between sounds in their own vocal repertoire. In addition to variations in the sounds (from noisy ones to those with a clear pitch), finches use the distribution of the spectral information, that is, the energy in the sound spectrum, to categorize the different types of vocalizations of other members of their species, in other words, not the pitch as such.[9]

As for melody perception, songbirds rely mostly on the spectral information and how that changes over time, or, more specifically, the changes in the spectral energy from one sound to the next. By contrast, humans listen to the pitch, paying little attention to the timbre.

One could say that songbirds listen to melodies the way humans listen to speech. In speech, humans focus mostly on the spectral information; this is what allows us to differentiate between the words "bath" and "bed." In music, melody and rhythm demand all the attention. Whereas in speech, pitch is secondary—it can say something about the identity of the speaker or the emotional significance of the utterance—in music, it is primary. This is an intriguing and as yet poorly understood distinction between the experiences of listening to music and listening to speech.[10]

A possible explanation, mentioned earlier, is that musicality is a by-product of cortical systems that were developed for speech and are supernormally stimulated by music. This would align with Steven Pinker's theory regarding the origins of musicality.

An opposing explanation, however, is also possible, namely that musicality *precedes* both language and music. In that case, musicality could be interpreted as a sensitivity that humans share with many nonhuman species, but in humans this predisposition has evolved into two partially overlapping cognitive systems: music and language.

Of course, this is all pure speculation. But I unexpectedly came across the beginnings of empirical evidence supporting this idea at an international conference in Austria.

Vienna, April 15, 2014. This morning, having just arrived at a conference on the evolution of language and music, I decide, by way of exception, to attend a lecture on a subject I know little about. The lecture is by Michelle Spierings, one of Carel ten Cate's new students.

At least five pages full of symbols, diagrams, and a plethora of exclamation marks: these are the notes I take during her lecture, which has the rather technical title "Prosodic Cue Weighting by Zebra Finches." I had only heard from Carel about this research in passing, possibly because it had little to do with rhythm, the subject we had been collaborating on for some time. But from the moment the study design is projected on the screen in the big conference hall, I am all ears.

The research examines how zebra finches learn to identify differences between sound sequences. In her presentation, Michelle calls them "syllables." The sounds consist of human utterances such as "mo," "ca," and "pu." The order of these speech sounds (syntax), as well as their pitch, duration, and dynamic range (spectral contour), is changed throughout a series of different behavioral experiments.

The zebra finches first learn the difference between the sequence *Xyxy* and *xxyY*, in which *x* and *y* stand for different speech sounds, and the capital letter for a musical accent: a bit higher, longer, or louder. For example: "MO-ca-mo-ca" as opposed to "mo-mo-ca-CA."

The finches then listen to an unfamiliar sequence, with altered accents and structure. The purpose is to test which aspect of the speech sounds the birds use to make the distinction: the musical accent or the order of the elements.

As Michelle shows, humans make these distinctions primarily on the basis of the order of the elements: *abab* is different from *aabb*, while *cdcd* resembles *abab*. Humans "generalize" the structure of *abab* to the as yet unheard *cdcd* sequence. This supports the idea that humans focus mainly on the syntax, or the order of the elements, when listening to such a sequence. Syntax (word order) constitutes an important characteristic of language (remember the earlier "man bites dog" example).

By contrast, the zebra finches turn out to focus mostly on the musical aspects of the sequences. This does not mean they are insensitive to the order (in fact, they were able to learn it to some degree), but it is mainly the differences in pitch (intonation), duration, and dynamic accents—the musical prosody—that they use to differentiate the sequences.[11]

This one short presentation completely changed how I look at zebra finches. Properly interpreted, the results presented could suggest that humans may share a form of musical listening with zebra finches, a form of listening in which attention is paid to the musical aspects of sound (musical prosody), not to the syntax and semantics that humans heed so closely in speech.

Once again, Darwin came to mind. Might the musical listening process of humans and zebra finches be more closely related than I thought?

To Listen like a Songbird

The research on starlings and zebra finches that I have described here reveals that songbirds use the entire sound spectrum to gather information. They appear to have a capacity for listening "relatively," that is, on the basis of the contours of the timbre, intonation, and dynamic range of the sound. This is a form of listening that had been observed earlier by music theoreticians and that led modern composers like Edgard Varèse, György Ligeti, and Kaija Saariaho to give timbre an important place in their compositions.[12]

Relative pitch in humans can mean more than just hearing relationships between pitches. Familiar melodies in which the pitch is rendered unrecognizable, for example, can also be identified from the contours of other aspects of sound. But humans are seldom interested in spectral contours.[13]

All of this raises intriguing questions: What is needed, for instance, for a human to be able to listen like a songbird? Or, conversely, is it possible for a songbird to listen to music the way humans do? More fascinating subjects for the research agenda.

Humans and songbirds thus have their own strategies and preferences when it comes to listening. While there is a clear difference in the attention they pay to pitch and timbre respectively, both species still have a certain flexibility when it comes to perceiving sound. Relative pitch turns out not to be limited to pitch alone.

On the train from Amsterdam to Leiden, May 28, 2013. By this time, three of the four zebra finches have completed their work and been moved back to the big aviaries. Only Bird 79 is still busy. Seated in the "quiet" train compartment, I leaf through the many tables and appendices that Carel sent me this week. Judging from the results as a whole, the zebra finches are capable of distinguishing between different rhythms. But—and this is the big surprise—they do not notice regularity in the stimuli.

This became apparent in the tests the birds completed after they had learned to distinguish between the regular and irregular rhythms. In this test phase, the zebra finches were exposed to new patterns, faster or slower versions of the earlier-learned sequences. They were not rewarded for their responses. Subsequently, their responses made it possible to determine whether they interpreted the newly heard rhythm as a go or a no-go stimulus.

It turned out the finches responded much less frequently when the normal go sound was speeded up or slowed down. They no longer appeared to recognize it, leading us to conclude that they probably hadn't noticed the regularity of the sound.

To investigate which characteristic of the learned rhythms the finches had used to make the distinction—after all, it was not the (for us) abundantly clear metronomic regularity of the go stimulus—we had constantly shortened the regular and irregular rhythms. The finches had performed each task equally well, regardless of whether the rhythms were long or short. We therefore had had to conclude that they did so by using information occurring somewhere near the beginning of the rhythm. Apparently the zebra finches remembered the absolute duration (the duration without reference to an external interval) of the first interval that differed in both rhythms, and used this information to successfully procure their food. Hence our conclusion that they did not detect on the basis of regularity.

Of course, we still do not know for a fact whether zebra finches are unable to differentiate on the basis of regularity. Even so, in this context, based on these stimuli, it appeared that regularity was at least not something that caught their attention. Metrical structure, so important to humans, appeared to be irrelevant to the finches.

Other research, however, revealed that, when a song by a member of their own species was lengthened or shortened in time, the zebra finches still recognized it as the same song.[14] This shows that, at least where their own song is concerned, zebra finches are sensitive to absolute pitch, timbre, and differences in loudness, and less so to absolute durations, which change significantly when a song is speeded up or slowed down. The latter suggests that zebra finches are particularly attentive to the relative rhythm of their song. This could mean that they hear regularity, but it is also possible that they focus more on the order of the song elements—does *B* follow *A*?—and less on the rhythm in which they are sung.

Nose Wheel

Our zebra finch study on rhythmic regularity was published in the journal *Behavioral Processes* in June 2015,[15] signifying the end of a period during which we had viewed beat perception primarily as a temporal phenomenon. If we had learned anything, it was that rhythmic structure is not the first thing zebra finches pay attention to. The evidence appeared increasingly to suggest that zebra finches focus primarily on intonation, timbre, and dynamic differences and minimally on the temporal aspects of sound. In fact, musical prosody might well be more informative for zebra finches than the temporal structure of the song elements. Yet another subject for the ever-expanding research agenda!

The results of the zebra finch study forced me to face facts again: what is obvious to humans is not necessarily obvious to animals. While I cannot help but hear regularity in regular rhythms, zebra finches appear to focus mainly on other "local" aspects, such as a single tone or time interval. This illustrates, once again, my favorite one-liner from the American psychologist James J. Gibson: "Events are perceivable but time is not."[16] The perception of time is only possible when something happens. In the case of zebra finches, this "event" seems to be the individual sounds to which they attribute certain characteristics and not so much the temporal structure of a sequence of sounds (the rhythm in which the sounds follow each other).

In this sense, humans listen more globally and abstractly, with greater attention to the whole. We are almost too good at seeing and hearing relationships, relationships which are often not there but have their source in our own experiences and expectations. This is why we find it surprising that other animals solve problems in ways seemingly much more complicated than our own. However, what is the simplest solution for us is not always the simplest solution for another animal species.

Consider this example of an unexpectedly simple solution for a difficult problem in the visual domain: to develop a search algorithm that can find photographs of airplanes on the internet. This is a difficult task because many of the countless possible photographs include depictions of objects closely resembling airplanes, such as birds or other white or metallic objects against a blue background.

The classic method in artificial intelligence would be to create a knowledge-based system that codifies precise rules (interpretable by a computer) about what does and does not constitute an airplane. The list could be quite long: an elongated and symmetrical object, two wings, a nose and a tail, small windows along both sides, a propeller on the nose or each wing, and so on. It is extremely challenging to compile a list of criteria that all airplanes would meet, but that would also allow airplanes to be distinguishable from, for example, birds and other airplane-like objects.

Recent computer simulations convincingly demonstrate that the most efficient way to determine whether an object in a photograph is or is not an airplane is, surprisingly, not to use a knowledge-based system. All the complicated reasoning turns out to be superfluous. The question—is there or is there not an airplane in the photograph?—can be answered much more simply and efficiently by focusing on one detail alone: is there or is there not a nose wheel in the photograph?[17]

Zebra finches and other animals that regularly take part in categorization experiments may be able to do just that. They listen, so to speak, to the "nose wheel" of the music: a detail that has little to do with the essence of the music. The bird remembers and recognizes one distinct detail, a detail that has resulted in food often enough to make it worthwhile for the bird to continue to focus on it.

What we know for sure is that humans, songbirds, pigeons, rats, and some fish (such as goldfish and carp) can easily distinguish between different melodies. It remains highly questionable, though, whether they do so in the same way as humans do, that is, by listening to the structural features of the music.[18]

A North American study using koi carp—a fish species that, like goldfish, hears better than most other fish—offers an unusual example. Carp are often called "hearing specialists" because of their good hearing. The sensitivity of a carp's hearing can be compared to the way sounds might be heard over a telephone line: though quality may be lacking in the higher and lower ranges, the carp will hear most of the sounds very clearly (see figure 7.1).

Three koi—Beauty, Oro, and Pepi—were housed in an aquarium at Harvard University's Rowland Institute, where they had already participated in a variety of other listening experiments. In the earlier experiments, they had learned they would receive food if they pressed a button at the bottom of the tank, but only if music was heard at the same time. The current experiment concentrated on the carps' music-distinguishing ability. As well as being taught to differentiate between two pieces of music (discrimination), Beauty, Oro, and Pepi were observed to see if they could recognize whether unfamiliar pieces of music resembled other compositions (categorization).[19]

In the discrimination experiment, the koi were exposed to compositions by Johann Sebastian Bach and the blues singer John Lee Hooker to see whether they could differentiate between the two. In the categorization experiment, the koi were tested to see if they could classify a composition as belonging to either the blues or the classical genre. In the latter experiment, they were alternately exposed to recordings of different blues singers and classical composers ranging from Vivaldi to Schubert.

The surprising outcome was that all three koi were able to distinguish not only between compositions by John Lee Hooker and Bach, but also between the blues and classical genres in general. The fish appeared to be able to generalize, to correctly classify a new, as yet unheard piece of music based on a previously learned distinction.

But what was the basis for the kois' decisions? How did they make the distinction? And what exactly did they listen to? If nothing else, the study clarified that they did not make the distinction based on the timbre of the music, because even when the classical and blues melodies were played on an instrument with a different timbre, the koi were still able to distinguish between them.

The koi research was inspired by a 1984 study describing the music-distinguishing ability of rock doves. It turned out that rock doves, too, can distinguish between compositions by Bach and Stravinsky. And, like carp, rock doves can also generalize what they have learned from only two pieces of music to other, unfamiliar pieces of music. They can even distinguish between compositions by contemporaries of Bach and Stravinsky.[20]

Rock doves and carp are therefore able to do something that is quite difficult for the average human listener: judge whether a piece of music was composed in Bach's time (the eighteenth century) or Stravinsky's (the twentieth century). Moreover, these species can do all of this with no significant listening experience, no extensive music collection, and no regular concert attendance. I suspect, therefore, that they perform the task on the basis of one distinct detail. This, in itself, is an exceptional trait. Yet it still offers no insight into the "perception, if not the enjoyment," of music. Although a successful tactic for rock doves and carp to generate food, it is still a far cry from musicality.

10 Rio and Ronan

Santa Cruz, June 3, 2014. Rio nudges her nose gently against the clenched fist of her carer. Waddling on her solid, finned feet, without increasing or reducing the pressure she applies to the fist in the slightest, she is led between several large water pumps to a shed clad in sheets of corrugated metal. The shed is, in fact, a fully equipped sound studio, replete with measuring equipment, speakers, soundproofing, and a floating floor. In this studio, marine mammals' hearing is tested on land. Special foot- and headrests have been installed so that the different marine mammals taking part in the experiments can rest comfortably and at the right height between the two loudspeakers. It is the so-called all-air facility.

I am visiting the Joseph M. Long Marine Laboratory, a research institute built on a rocky terrace at the foot of the Coastal Science Campus of the University of California, Santa Cruz. Colleen Reichmuth has invited me to accompany her and her pinniped-specialized research group for a day. Dozens of spotted seals, walruses, sea lions, and harbor seals live and work here.

Colleen's small research team of biologists and physiologists is assisted by several dedicated volunteers, who work at the facility together with the researchers all week. The team performs exercises and behavioral or listening experiments most mornings. Each animal has a permanent carer who regularly exercises along with it: raise the foot, roll onto the back, growl briefly, open the mouth in response to a gesture or whistle signal. In this way, each animal develops a repertoire of gestures and actions for possible future use as the building blocks of new behavioral experiments. The exercises are also convenient for maneuvering the animals into a position that will allow the veterinarians to do their work when a medical examination is required. This is no luxury, given that most of the animals here are far from lightweights.

The large facility has a superb view of Monterey Bay. The massive skeleton of a blue whale stands on display in the parking lot, surrounded by shipyard-like buildings covered in sheets of corrugated metal. The light-blue pools dotting the site are filled with seawater, which is pumped out of the bay by a series of installations and then brought to the appropriate temperature for the different animals. The ringed seal from Alaska is accustomed to colder temperatures than the Hawaiian monk seal two pools farther down.

Rio, who has just waddled over, also has access to various saltwater pools and "haul-out areas" (where pinnipeds temporarily leave the water to rest, molt, or breed). Born in captivity but neglected by her biological mother, Rio was raised by one of the laboratory's employees. She made her name through a study examining long-term memory in sea lions. Though it had already been known for some time that sea lions can recognize their breeding and feeding grounds, as well as individual members of their species, the now-classic study demonstrated that Rio still had ready access to knowledge of similarities and differences between images of numbers and letters that she had acquired ten years earlier.[1]

Today Rio is taking part in a short demonstration of how pinniped hearing is tested. The screen in the observation room shows her looking intently at the loudspeakers, waiting to hear something. A technician then plays several soft sounds over the loudspeakers and measures her reaction. Rio has learned to press her nose against a switch when she hears a sound. For this, she is rewarded with a herring at the end of each session.

The facility also conducts all kinds of underwater tests. On this occasion the team performs a hearing test in the cold pool with Nayak, a ringed seal born in Alaska in the wild, by means of a specially constructed underwater installation of interconnected white PVC pipes. After hearing a short whistle signal from her trainer, Nayak plunges into the pool and comes to a standstill in front of the PVC crossbar attached to the edge of the pool just under the surface of the water. She has learned to lie perfectly still, pay attention to what happens, and, depending on what she hears through the underwater loudspeakers, press a button. By changing the tone in the frequency and the sound pressure level, the researchers are able to learn what Nayak does and does not hear.

Because she has to perform well in a number of tests to get a reward, the experiment has been designed to offer almost as many easy as difficult sound samples. The level of difficulty is monitored and either increased or decreased depending on how well she performs. Although this complicates the results analysis somewhat (more complex statistics are required, for example, than with a go/no-go paradigm), a regular reward keeps the game entertaining and challenging for Nayak.

The psychophysical research at the Long Marine Laboratory focuses on investigating the hearing range and sensitivity of several pinniped species. Behavioral changes—such as "running away," being startled, or losing concentration—when confronted with loud or unexpected noises are also observed. Colleen's research group is mapping out the possible impact of sound on the ecological environment (such as Alaska or the Pacific Ocean) of these marine mammals. Today the same marine environments are largely dominated by oil platforms, ships, and explosions, the effects of which are scarcely known. The laboratory's income thus comes in part from government and the oil industry.

When the experiment ends, Nayak is tossed a frozen fish encased in a thick layer of ice to reduce the boredom. This keeps her happy in the round pool for at least thirty minutes.

Sprouts, a harbor seal from local waters, lives next to Nayak and has been here for almost as long as Rio. Whereas Rio has small, leaf-shaped ears, Sprouts only has two little indentations in his head. Beautifully spotted, he is covered with short, unidirectional hairs growing out of a solid layer of blubber and has an elegantly rounded head set on a thick neck. Not only is Sprouts bigger than Rio, he also has different eating habits. This is immediately visible when he is offered a herring, which he dispatches swiftly and noisily. He catches the herring and severs it in two with great force just behind the gills.

Sprouts's unique talent is his ability to make guttural sounds on command. He can produce a beautifully resonating growl. Having learned a small repertoire from Ron Schusterman, the laboratory's founder, Sprouts is one of the first mammals shown to have limited vocal learning.[2]

With a quick flick of her wrist, Colleen gets Sprouts to roll over onto his back, then asks me to put my hand on his throat. I lay my hand on his thick neck and feel the larynx vibrating vigorously. Sprouts is demonstrating his singing talent. My initial assessment: a full bass.

Amsterdam, April 3, 2013 (one year earlier). A science journalist friend of mine surprises me this morning with an intriguing article about a sea lion that moves to the beat of the music. "Look, another musical animal! What do you think? I'd like to hear your view," he writes in an e-mail.

A few days earlier, the article had been made available under embargo to the press only, a strategy adopted by an increasing number of scientific journals to maximize the amount of press attention given to an exceptional or mediagenic study. While in itself a sound strategy for maintaining good contact with the press, the embargo also ensures that a potentially exceptional study is read both carefully and critically rather than ending up on a heap.

I start to read the article immediately and quickly realize it is not just any study. The article describes a series of carefully constructed experiments in which a California sea lion called Ronan bobs her head up and down to the beat of different pieces of music, each of which is played at different speeds: she is, in effect, what could be called a "headbanging" sea lion.

That the study involved a sea lion is interesting for at least two reasons. First, sea lions are not known to have vocal learning. Like humans, they have a larynx. In the phylogenetic sense, this is an extremely old vocal system found only in vertebrate animals. However, to date, there is still no evidence that sea lions' vocalizations resemble a learned repertoire of sounds.

It is difficult to prove the absence of a trait, in this case the undoubtedly as yet undiscovered vocal learning ability of sea lions. However, there are many indications that sea lions do not have this ability, unlike other members of the pinniped family, such as seals and walruses. Following this line of reasoning, the study with Ronan can be seen as a refutation of the VL hypothesis. And this is exactly what the article's authors proposed. In subsequent months, the excitement and unrest grew among colleagues. Was this or was this not a falsification (or negation) of the VL hypothesis? I decided to go and observe with my own eyes.[3]

Santa Cruz, June 3, 2014. It is exactly thirty years ago to the day that I first came to the United States. I had been fascinated with musical computers, and that fascination had taken me straight to the institution pioneering this area of study at the time, Stanford University in Palo Alto. Today, though, it is musical animals that have my attention. I look forward to meeting Ronan, the California sea lion who had created such a stir the year before. I have traveled to California especially to see her.

Peter Cook, who did the beat perception experiments with Ronan the results of which were published last year, had written to me a few weeks earlier to say that this type of experiment with Ronan had ceased and he had moved to another university. As a result, I do not expect to see Ronan in action today.

I am therefore pleasantly surprised when Colleen tells me that she has been preparing for my visit over the past few days and will demonstrate the experiment again live. This is impressive. A year is a long time, and Ronan has since participated in a wide range of other experiments. But Colleen has decided to demonstrate Ronan's beat perception especially for me and to repeat the experiment exactly as it was done a year ago.

Like an Olympic star, Ronan waddles into the haul-off area enclosed with a wire-mesh fence, her nose pressed as usual against her trainer's fist. The session begins with the daily routine. On a short whistle signal, Ronan rolls onto her back, allows her flippers to be inspected, then rolls over again. On a second whistle signal, she extends her neck and opens her mouth for dental inspection. She then calmly carries out a series of activities in the requested random order, each, of course, for a piece of herring. By now, the bucket of herring is half empty.

Ronan's current trainer, Andrew Rouse, is responsible for the logistics. Next to a small pool farther up, he has assembled all the equipment necessary for the special demonstration. In the meantime, an excited Ronan is already doing laps in the pool. Colleen and I sit on a bench and peer through the wire mesh to see what will happen next.

Ronan cannot see her trainer. The wire mesh is covered with a sheet of plywood containing a small hole for a PVC tube. Herring will be offered regularly through the hole as a reward. In front of a small, raised barrier on the other side of the sheet of plywood, Ronan is waiting attentively with her neck extended.

A metronome-like tick is played over a loudspeaker. Ronan immediately begins headbanging to the rhythm of the ticks. After about ten ticks, the noise stops, and she is given a herring. The speed of the metronome is then accelerated slightly. Ronan catches on after the second tick already and continues to move her head exactly in time with the ticks. She is rewarded with a second piece of herring.

Then comes a song by Earth, Wind & Fire: "Boogie Wonderland," the classic 1979 disco hit. It takes Ronan less than half a second to figure out the tempo and move in perfect time to it. She replicates the tempo faster than I am able to with my foot. It makes no difference at all to her whether the music is played more quickly or slowly. She identifies the beat immediately, earning one piece of herring after another.

I am touched by her zeal and the apparent pleasure she takes in moving to the beat of the music. Her pleasure resembles that of Snowball the cockatoo, who could headbang to "Everybody" by the Backstreet Boys. Ronan, however, is doing it all on her own, without seeing her trainer. And unlike Snowball's 15 percent accuracy, Ronan moves exactly in time to the beat for most of the experiment.

I had already been convinced by the article and the analyses suggesting that none of this behavior was coincidental. But watching Ronan perform it in the flesh is even more exciting, if for no other reason than that she is still able to do the experiment, which she had last participated in more than a year ago, with great pleasure and precision.

Falsification?

During lunch with Colleen and several members of her research group at a nearby Greek restaurant, I broach the subject of vocal learning in mammals and the pressing question of whether Ronan's behavior indeed represents a falsification of the VL hypothesis. Colleen refers to the best-known exception among the pinnipeds: Hoover. Hoover was a seal with a small but impressive repertoire of imitated utterances, such as "Well, hello deah, how are ya?" and "Get outta here!" all with a heavy New England accent. He had been found abandoned on a beach in Maine at a young age and raised by a couple in their home. Eventually Hoover moved to the New England Aquarium in Boston, where he died in 1985. His offspring—he had six of them—never displayed his talent, despite attempts by volunteers in the rescue center to cultivate it.

Was Hoover therefore an exception? Colleen thinks not. In the past, Ron Schusterman had tried to teach new sounds to Sprouts, the laboratory's harbor seal, and he had succeeded. Sprouts had not learned to say complete sentences, the way Hoover had, but after a lot of practice, he had mastered a small repertoire of short, distinct sounds.[4]

No one had ever demonstrated that sea lions have vocal learning, apart from the sounds they make in the wild, namely, a limited collection of growling, coughing, puffing, and snoring noises. Colleen explains that the role of vocalization is mostly important for communication between mothers and their pups. Every mother has her own call, and a pup can already recognize it after one or two weeks. This is, of course, a far cry from what could be called vocal learning, which involves the capacity to learn new sounds. Sea lions simply do not have the neuronal and muscular adaptations necessary to make flexible vocalizations, something seals and walruses are theoretically capable of. Genetically speaking, therefore, sea lions and seals are relatively far removed from each other. Their common ancestor lived some twenty million years ago. By way of comparison, the common ancestor of humans and the great apes lived roughly five to ten million years ago.[5]

In short, until someone can demonstrate that sea lions have vocal learning, based on the lack of evidence and the distant relationship with animals that do have that ability, we must assume that sea lions do not have it. Yet they can perceive the beat and move in time to the beat of music. In principle, therefore, the study with Ronan is a falsification of the VL hypothesis.

It is wise, however, to remain cautious about the scope of such a conclusion. To date, we simply do not know precisely which animals have vocal learning and which have beat perception. Ronan could also be an exception, or perhaps her vocal learning ability has yet to be discovered. Nor can we answer the question yet of whether vocal learning is essential or just one of various possible preconditions for beat perception. It is therefore high time to do a sample calculation to ascertain how likely it is that vocal learning might be related to beat perception.

Sample Calculation

If we restrict ourselves to mammals, birds, and reptiles, then there are an estimated 25,000 animal species in total. Of them, probably 7,100 have vocal learning, namely, 300 hummingbird species, 400 psittacines, 5,000 songbirds, 1,300 bat species, 100 cetaceans, and a few other isolated species, including humans. Of the five studies published to date on animal species without vocal learning, only one has irrefutably demonstrated beat perception: the study on Ronan. This gives us a ratio of 1:5. Multiplying that by a rough estimate of the number of animal species with and without vocal learning, respectively, gives a relative probability, or likelihood, of 1:2.[6] In other words, beat perception is twice as likely to be found in an animal with vocal learning as it is to be found in an animal without it. This lends support, therefore, to the VL hypothesis, provided it is enhanced with an estimation of probability.

One could also be less stringent by counting the doubtful cases too, such as the studies involving chimpanzees and bonobos. The ratio would then be 4:5, of which only the study with rhesus macaques that I conducted with Hugo Merchant could be seen as evidence in support of the VL hypothesis. In that case, the relative probability would be roughly 2:1, in other words, exactly the opposite. Proof, therefore, of the relative improbability of the VL hypothesis.

This is, of course, a simplified calculation.[7] For example, only a limited group of animal species is known for certain to have vocal learning—roughly 100, rather than the estimated 7,100. All kinds of genetic and phylogenetic characteristics also make a uniform probability of occurrence of animal species with or without vocal learning an overly simplistic assumption. Even from the most conservative perspective, however, given what we know to date, the study with Ronan can be seen as a compelling, if not definitive, counterexample to the VL hypothesis. But given what we do not yet know (and in this case, that is a lot), it could well be precisely the opposite. I think it is still premature to ask Tecumseh Fitch to return the bottle of Rioja.

The sea lion study had a major impact on the small group of international researchers who until then had mostly supported the VL hypothesis. The idea of vocal learning as an absolute precondition for beat perception (a key characteristic of musicality) was no longer tenable or, at the very least, had to be nuanced.

Erich Jarvis, a research pioneer in the genetic underpinning of vocal learning, wrote an article in 2012 suggesting that vocal learning developed gradually: it was therefore not something you had or did not have.[8] In 2014, Ani Patel presented an alternative for his vocal learning hypothesis, suggesting that vocal learning was no longer an essential precondition for beat perception.[9] The relationship between the medial premotor cortex (MPC) and the primary auditory cortex (A1) was central to his new proposal. This was similar to the ideas that Hugo Merchant and I had developed in 2014 based on the recent literature studying rhythm perception in primates (see figure 4.1).[10] When different researchers present comparable ideas simultaneously, it is usually a sign that scientific consensus is not far off.

Santa Cruz, June 3, 2014. One last ritual has been planned to mark the end of my visit to the Long Marine Laboratory. Late in the afternoon, Ronan is led from the adjacent pool to the viewing area, where I had witnessed the head-banging experiment. She waddles straight over and sits down next to me. At a signal from Colleen, I bring my fist toward her. She immediately prods my fist gently. Colleen then instructs me to bring my hand slowly to my face.

Ronan knows the routine. I feel a slight pressure on my cheek and warmth emanating from her nostrils. She is giving me a farewell kiss. A gentle, lingering kiss that lasts until the whistle blows. I know it is not what it seems, but it makes me blush all the same.

Afterword

Amsterdam, October 10, 2013. A woman in a white coat bends over me and asks if I know where I am and what day it is. I have no idea. I look around me and glimpse beds on high legs in a sun-drenched room and computer screens hanging upside down from the ceiling but far too high up. Strange. Is this a hospital? A short time later someone else comes and asks me exactly the same questions. Why the persistence? Why the questions? And the day? Well, I had resolved to remember that.

I look at the clock. Suddenly it is midnight. I am lying in a dimly lit room—in a hospital, that much I know by now. A handwritten note lies on the bedside table next to me. It contains answers to my questions, which I had repeatedly asked in the same order. How do they know that? I start to become agitated and, above all, suspicious. I set the note aside and immediately forget what it says. When the dizziness passes, I pick it up again and am surprised to read, for a second time, the same answers to the same questions. How could this be? Something must be terribly wrong.

It is Tuesday. It seems I was knocked off my bicycle and landed on my head. At least that is what the note says. I had been heading to one of my regular local cafés to write and have a bite to eat. A severe concussion and something broken or bruised in my back, they say. Now I begin to worry. Might my head have been injured? As for my back, that much is clear, because I feel a sharp pain there. That will pass. But my head?

Occasionally, someone comes and shines a flashlight in my eyes and asks the now-familiar questions: Where am I? What day is it? I practice the answers because it seems important, but they evaporate again instantly. Earlier in the day my girlfriend, tiring of my endlessly repeated questions, had written them down on a piece of paper with the telling heading "FAQs," and laid it on the bedside table.

I had just received a generous grant that would allow me to spend a year at the Netherlands Institute for Advanced Study in the Humanities and Social Sciences (NIAS) in Wassenaar and work full-time on this book and a number of other projects, including organizing an international workshop on musicality to be held at the Lorentz Center for the Sciences in Leiden. This would not be happening now, at least not as I had imagined it. During the first few weeks in Wassenaar, it took an enormous effort for me to walk even the shortest distance in the nearby dunes. Someone always had to accompany me because I lost my way after just one or two bends in the path. I sleep endlessly and worry mostly about whether my head will ever recover.

Wassenaar, December 3, 2013. I leaf through my notes from the past few weeks and keep coming across the same ideas formulated almost identically. Am I forgetting more than I realize?

Suspiciously, I reread several texts. Some passages appear to have been written by someone else. Access to my own memory has clearly changed. A week ago, my neurologist had assured me that this was normal. All kinds of things could be tested but my brain was obviously still recovering. A test would only be a snapshot, he had commented strategically. Recovering? It had already been two months since the accident. What if my memory doesn't improve?

My unease is fueled by the fact that I have twice had to postpone a lecture series for the Universiteit van Nederland (a platform where leading Dutch scientists give free lectures on the internet). Though I am still rehearsing my text, today as well, I keep forgetting what comes next in the narrative. I have no overview. I can concentrate on the details but am hopeless with the broader picture. This suddenly reminds me of zebra finches. Is this what it feels like to be a zebra finch?

After walking around my room in circles for the umpteenth time, practicing the lecture and finally acknowledging that things aren't going much better than yesterday, I give up and decide to cancel all my speaking engagements for the coming months. I feel relieved and sad at the same time.

Wassenaar, February 4, 2014. In about five weeks' time, twenty-three researchers from around the world will be coming to Leiden for the Lorentz workshop to talk about musicality for a week. At the moment, I cannot imagine I will be able to participate at their level. It is taking me forever to plan the workshop, to think through all the details one by one—the program, the working sessions, the research questions. But the workshop must go ahead, if necessary, without me.

Today Carel ten Cate is coming to talk through the program again and check that I haven't overlooked anything. We draw up a strict regimen for the rapporteurs. They will record all the afternoon sessions, then report back to the others at the end of the day on what was discussed and proposed during the working sessions. Two students will also take notes on each of the other activities so that I don't have to worry about my shaky memory. The

Figure 11.1
Participants at the Lorentz workshop. *Front row (seated, from left to right)*: Isabelle Peretz, W. Tecumseh Fitch, Ani Patel, Björn Merker, Henkjan Honing, Iain Morley, and Sandra E. Trehub; *behind (standing, from left to right)*: Carel ten Cate, Simon E. Fisher, Willem Zuidema, Yukiko Kikuchi, David Huron (white shirt), Hugo Merchant (checked shirt), Laurel J. Trainor, Martin Rohrmeier, Judith Becker (white cardigan), Marisa Hoeschele, Jessica Grahn, Yuko Hattori (white blouse), Bruno Gingras, and Geraint Wiggins.

plan is to draw up a research agenda for the coming years by the end of the conference.

In the end, the workshop went much better than I dared to hope. From the first day already, the atmosphere within the group of researchers was highly constructive, partly due to the careful selection of participants. I had discussed the selection thoroughly with Sandra Trehub, Isabelle Peretz, and Carel ten Cate. It wasn't only about bringing together the most interesting researchers and their complementary areas of expertise, preferably with the right balance of junior/senior and male/female. An even more important criterion was to assemble participants who were prepared to think freely and share their thoughts openly in an interdisciplinary context. The result was that none of the intransigent positions I have often experienced at expert workshops emerged. In particular, the motivation to focus on what we didn't know, rather than on trying to convince colleagues about our own research and insights, resulted in a highly productive week. The theme of the first morning was the evolutionary origins of musicality.

History of Musicality

Although, by now, countless studies had shown that the prehistory of music and the evolutionary origins of musicality had biological foundations, opinions at the workshop varied widely about the latter. Roughly speaking, there were two lines of thought. One group of researchers argued that music was primarily a cultural phenomenon, because, in their opinion, most musical knowledge and skills were learned. Moreover, many hours of study were required to master a voice or an instrument, unlike speech, for example. These researchers further contended that music was such a recent phenomenon that it couldn't possibly have influenced the form and structure of human cognition or biology, let alone the evolution of either.[1]

Another group of researchers claimed exactly the opposite. In their view, music had a long history that was deeply rooted in our biology. They referred to archaeological finds of musical instruments, such as a 45,000-year-old bone flute. This instrument, with holes specifically spaced to allow melodies to be played, almost certainly derived from a long history of musical activity. Music, they suggested, might even be one of the oldest human cognitive functions.[2]

A third theory, somewhere in between these two positions, was that music was a transformative technology: a human invention based on existing mental functions (such as emotions, memory, language, and speech) that had slowly and irreversibly changed our lives over tens of thousands of years.[3] It could be compared with the mastery of fire. No natural selection or gene was responsible for human development of that skill, yet it had a huge impact on our lives, not only on our cultural and social structure but also on our biology. A similar line of reasoning could apply to musicality.[4]

Over time, a variety of alternatives had been suggested for each of these lines of thought, by thinkers ranging from Rameau and Rousseau, Spencer and Darwin, to Stumpf, Mithen, and Dissanayake, all of whom considered human musical expression to be so exceptional and profound that they could not conceive of there being no evolutionary component. Given the importance of music in our society, musicality must be an evolutionary adaptation.

Of course the problem was that, given the absence of definitive empirical evidence, all these theories were no more than speculations. If playing the flute or drumming had been an essential activity for survival or social cohesion, we might have found anatomical adaptations in fossilized human remains. Unfortunately, the musical brain and musical sounds do not fossilize. So we have had to make do with the few remaining indirect traces of prehistoric musical activity, such as a lone flute that has withstood the ravages of time. Even then, a period of 45,000 years means nothing in evolutionary terms. The same bone flute might just as easily be seen as a rather recent cultural artifact that has little to do with our biology. Most of the evolutionary theories about the origins of music therefore appear doomed to remain speculative, stories that cannot (or can only partially) be substantiated—or falsified, for that matter—with facts. Scientifically speaking, in other words, these theories represent a dead-end street.

As a result, one of the most important conclusions of the Lorentz workshop was that if we ever wanted to be able to say something about the evolution of music and musicality, we would first have to establish what the components of music were and how we could demonstrate their presence in animals, including humans (see figure P.1). Much could be learned about these components, not only through interdisciplinary and comparative research, as described in this book, but also with the help of methods from, for example, developmental psychology, neurobiology, and the rapidly developing discipline of genetics.[5] During the remainder of the workshop, we therefore focused on methods and techniques that could tell us something about musicality in the here and now. For the time being, we decided to leave the debate about the evolutionary history of music for what it was.[6]

Amsterdam, April 16, 2014. This morning I receive an e-mail from the senior commissioning editor of the Royal Society's *Philosophical Transactions B*, the oldest scientific journal in the world (Charles Darwin had published in it), inviting me to prepare a special issue of the journal—to mark its 350th anniversary—dedicated to the theme of musicality. The opportunity to put musicality on the international research agenda is suddenly very close. I am delighted. One of the workshop participants had informed the journal's editorial board of our previous week's gathering. All the fellow researchers who had helped draw up the research agenda were on more or less the same page: there were no disproportionate egos, no theories competing for the truth, as often seems to be the case in other scientific disciplines. Interdisciplinary collaboration on the subject of musicality appears to be more plausible, perhaps because positions have not yet been adopted and theories have not yet become entrenched. However, perhaps everyone also realizes that musicality is not a monolithic concept but rather a composite encompassing many different components for which contributions from different disciplines are essential.

We still have much to discover about musicality. This makes researchers happy. Me, at any rate. Realizing what you do *not* know is infinitely more informative than realizing what you *do* know. In this sense, there is an important difference between knowledge and research. My slowly healing concussion reinforced that feeling. What I knew was sometimes not available to me, at least not concretely. Fortunately, though, I was still able to repeatedly recall what the interesting questions were, as I discovered while writing this book. My notes from the previous months revealed that I had remembered them, although others had undoubtedly helped me in that process. But it enhanced my pleasure, "if not … enjoyment," in mapping out the genealogy of musicality.[7]

Summary

Charles Darwin assumed that all animals can detect and appreciate melody and rhythm simply because they have a nervous system comparable to that of humans. He therefore had no doubt that human musicality had a biological foundation and a long evolutionary history. This is the underlying assumption of this book, in which I search for answers to the question of what makes us musical.

My journey began in 2009 when, together with a group of Hungarian researchers, I discovered that human infants have beat perception. Infants hear regularity and notice irregularities in a varying rhythm. Beat perception is a prerequisite for being able to dance or make music together. Both the brains and the hearing capacity of human infants turned out to be primed for music. This conclusion contributed to my decision to write a book about musicality (*Iedereen is muzikaal: Wat we weten over het luisteren naar muziek*; published in English as *Musical Cognition: A Science of Listening*), but also raised the question of whether musicality was exclusively human or perhaps a trait possessed by all animals, as Darwin suspected.

To answer this question, I approached numerous behavioral biologists and neurobiologists. At the time, biology was still an unknown discipline for me. I turned to neurobiology first, to determine whether it was possible to perform the same listening experiment with rhesus macaques as we had done with human newborns. In this way, I hoped to be able to discover whether rhesus macaques, whose ancestors had split off from other non-human primates at about the same time as our human ancestors—some twenty-three million years ago—also respond to the beat of music.

Most of the primatologists I spoke with were skeptical about the idea of using a measurement method on monkeys that had once been used on human infants, namely, an EEG (involving electrodes being attached to the skull). But the neurobiologist Hugo Merchant of the Universidad Nacional Autónoma de México leapt at the idea. Within a few weeks, his research group had assembled all the necessary software and hardware to make the first trial measurements. Unfortunately, after more than a year of setbacks and new insights, we had had to conclude that rhesus macaques do not have beat perception. This finding ran totally contrary to what I had expected and to Darwin's assumption.

Until then, scientists had widely believed that the heartbeat was the source of beat perception. After all, all mammals, including rhesus macaques and humans, hear their mother's heartbeat in the womb. Our research findings made that hypothesis much less likely. Apparently, specific neural networks are necessary to enable beat perception. In rhesus macaques, these networks are weaker or perhaps even nonexistent. The brain, therefore, rather than the physiology is the decisive factor here.

Our study and others led to the development of the "gradual audiomotor evolution" hypothesis, which suggests that the neural networks allowing for beat perception in humans are less developed in the great apes and altogether absent in rhesus macaques. The hypothesis also predicts that chimpanzees must have a rudimentary form of beat perception. This theory led me to visit the Primate Research Institute in Inuyama, Japan, where the primatologist Yuko Hattori was investigating the musicality of chimpanzees. Her initial findings suggested that chimpanzees do have beat perception, thus allowing us to date the origins of human beat perception to the common ancestor of humans and chimpanzees, some five to ten million years ago.

In 2009, the same year that our study on human newborns was published, a groundbreaking article appeared about the sulfur-crested cockatoo Snowball, who could dance to the beat of music. He appeared to adapt his tempo when the song was played faster or slower. Snowball was responsible for a renewed global interest in the biological foundations of musicality and represented a first major step forward in support of Darwin's hypothesis.

This interesting research result led in turn to more new questions, such as: why did humans and cockatoos (our common ancestor lived about 320 million years ago) but not rhesus macaques develop beat perception? The ability to hear regularity in music may be based on vocal learning, that is, the ability to learn and imitate new sounds. The "vocal learning as precondition for beat perception" (VL) hypothesis predicts that, rather than sharing beat perception with other mammals, such as horses, dogs, and rhesus macaques, humans may share it with specific bird species, such as cockatoos, budgerigars (common parakeets), and zebra finches.

For this reason, in 2012, as well as performing listening experiments with rhesus macaques, I began investigating beat perception in zebra finches in close collaboration with Carel ten Cate of Leiden University. Carel showed me the way in behavioral biology. Zebra finches are a preferred animal model for research on the role of genetics and the environment in evolutionary processes. They had already proved to be extremely useful for research on the evolution of language and speech, and now turned out to be equally effective for research on the evolution of musicality.

Zebra finches have vocal learning. They learn their characteristic song from their father or other family members. After two years and many listening experiments, though, we had to conclude that zebra finches pay no attention to regularity in rhythm. Vocal learning is not, therefore, the only precondition for beat perception. As a result, the VL hypothesis had to be adapted. It was also becoming increasingly apparent that zebra finches focus more on intonation, timbre, and dynamic differences than on the temporal aspects of sound. This "musical prosody" might be more informative for zebra finches than the rhythm of the song elements, providing yet another new subject for the ever-expanding research agenda.

In 2013, an extensive study from the laboratory of the behavioral biologist Colleen Reichmuth of the University of California, Santa Cruz, unexpectedly appeared and further fueled the debate about the VL hypothesis. The publication compellingly demonstrated that Ronan, a California sea lion, had beat perception, whereas this species was widely believed not to have it. Although it is difficult to prove the absence of a trait—in this case the undoubtedly as yet undiscovered vocal learning ability of sea lions—there were many indications that they did not have this ability. Yet Ronan appeared to be able to move convincingly to the beat of the music. This finding further undermined the VL hypothesis. The scientific literature went on to emphasize the social role of beat perception and its link to "prosocial behavior" in humans and other mammals, that is, behavior aimed at promoting the welfare and well-being of others.

In addition to beat perception, relative pitch (the ability to recognize a melody whether it is played higher, lower, faster, or slower) also turns out to be a fundamental building block of musicality. Most animals have perfect pitch. They remember and recognize sounds on the basis of absolute frequency (the vibrational frequency), not on the basis of melodic progression or interval structure, as humans do.

Yet perfect pitch has little to do with musicality. Relative pitch, on the other hand, does. It is a trait that has been well researched in humans. Neuroscientific studies reveal that relative pitch uses a complex network of different neural mechanisms, including interactions between the auditory and parietal cortices, the latter of which are involved in sensory and cognitive functions such as attention and spatial orientation. This network appears to be lacking in the brains of songbirds, for example. In the context of research on the biological origins of human musicality, these findings make the question of whether humans share relative pitch with other animal species all the more fascinating.

Regarding the perception of melodies, songbirds rely mostly on the changing timbre of a song. By contrast, humans listen to the pitch, paying little attention to the timbre. One could say that songbirds listen to melodies the way humans listen to speech. In speech, humans focus mainly on the changing timbre and less on the pitch itself. This is what allows us to differentiate between the words "bath" and "bed." In music, it is melody and rhythm that demand all the attention.

Compared with songbirds, humans listen more globally and abstractly, with greater attention to the whole. We hear relationships between the notes, relationships that have their source in our own experiences and expectations and not in the notes themselves. Ultimately, though, I believe Darwin will, for the most part, be proven right, and that research will show humans and animals to have comparable sensitivities in the way they listen to melody and rhythm. The challenge is to design the right experiments—experiments that are adapted to the animals' living conditions—through which to investigate musical listening in animals.

Thanks to a generous grant, I was able to work on a gradually emerging research agenda in 2014 and organize an international workshop. A group of twenty-three researchers came to the Netherlands from around the world to talk about musicality for a week. From day one, we agreed that musicality is not a monolithic concept but a composite encompassing many different components for which contributions from different disciplines are essential. As so, after it quickly became apparent that musicality remains a subject about which there is still much to be discovered, we drew up a research agenda for the coming years (published as *The Origins of Musicality* in 2018).

Acknowledgments

I am very grateful to the Netherlands Institute for Advanced Study in the Humanities and Social Sciences (NIAS) in Wassenaar and the Lorentz Center in Leiden, which gave me the opportunity to spend an entire academic year reading about the relationship between music, musicality, and biology; to reflect on it; and subsequently to organize an international workshop on the subject. Although the year passed faster than I would have liked, owing to a bicycle accident and its aftermath, it provided me with the mental space to envisage a first, provisional bridge between music cognition and biology, and to describe the initial exploratory research steps in what would later become the fabric of this book.

Many people supported me during the writing process. I am indebted to Pieter de Bruijn Kops, Henk ter Borg, Carel ten Cate, Tijs Goldschmidt, Pim Levelt, Alice ter Meulen, and Anne-Marie Vervelde for their valuable advice. Special thanks go to Bob Prior and Anne-Marie Bono at the MIT Press for their enthusiasm and support, Sherry Macdonald for the beautiful translation, and the Dutch Foundation for Literature for making the translation possible. I also thank Peter van Gorsel, Petry Kievit, René van Veen, and many others who commented on the Dutch manuscript (or parts of it) at different stages. My thanks go as well to Merwin Olthof, Carola Werner, Iza Korsmit, and Jeannette van Ditzhuijzen for their help in the planning of teaching and research activities, and other logistical matters while I was writing this book. In the period before and after my stay in Wassenaar, tables at the cafés Captein & Co., Mads, and Spanjer & Van Twist in Amsterdam often afforded me precious writing opportunities. Finally, I would like to sincerely thank my research group, a diverse assemblage of distinct young minds with various talents and ambitions. Time after time they showed me that, as well as creating considerable confusion and uncertainty, interdisciplinarity can generate valuable and much-needed insights and solutions.

Explanatory Note

Logbook excerpts that appeared previously in *Op zoek naar wat ons muzikale dieren maakt* (Nieuw Amsterdam, 2012), which was published to mark the twelfth Van Foreest Public Lecture in 2012,[1] have been incorporated in the chapters "Shaved Ear," "Mirroring," and "Beat Deaf." Part of "Perfect Pitch" was published earlier in *Hoe zwaar is licht* (Balans, 2017; in Dutch only).

Notes

Preface

1. C. Darwin, *The Descent of Man, and Selection in Relation to Sex*, vol. 2, 1st ed. (London: John Murray, 1871), 333.

2. H. Honing, *Musical Cognition: A Science of Listening*, 2nd ed. (New Brunswick, NJ: Transaction Publishers, 2013); originally published in Dutch as *Iedereen is muzikaal: Wat we weten over het luisteren naar muziek*, 2nd ed. (Amsterdam: Nieuw Amsterdam, 2012).

3. A. Lomax and N. Berkowitz, "The Evolutionary Taxonomy of Culture," *Science* 177, no. 4045 (1972): 228–239, http://doi.org/10.1126/science.177.4045.228.

4. B. Nettl, "An Ethnomusicologist Contemplates Universals in Musical Sound and Musical Culture," in *The Origins of Music*, ed. N. L. Wallin, B. Merker, and S. Brown (Cambridge, MA: MIT Press, 2000), 463–472.

5. P. E. Savage, S. Brown, E. Sakai, and T. E. Currie, "Statistical Universals Reveal the Structures and Functions of Human Music," *Proceedings of the National Academy of Sciences* 112, no. 29 (2015): 8987–8992, http://doi.org/10.1073/pnas.1414495112.

6. Honing, *Musical Cognition*.

7. Darwin, *The Descent of Man*, vol. 2, 333.

1 Shaved Ear

1. H. Honing, "Without It No Music: Beat Induction as a Fundamental Musical Trait," *Annals of the New York Academy of Sciences* 1252, no. 1 (2012): 85–91, http://doi.org/10.1111/j.1749-6632.2011.06402.x.

2. B. Tarr, J. Launay, and R. I. M. Dunbar, "Music and Social Bonding: 'Self–Other' Merging and Neurohormonal Mechanisms," *Frontiers in Psychology* 5 (2014): 1–10, http://doi.org/10.3389/fpsyg.2014.01096.

3. M. Velliste, S. Perel, M. C. Spalding, A. S. Whitford, and A. B. Schwartz, "Cortical Control of a Robotic Arm for Self-Feeding," *Nature* 453 (2008): 1098–1101, http://doi.org/10.1038/nature06996.

4. R. Näätänen, P. Paavilainen, T. Rinne, and K. Alho, "The Mismatch Negativity (MMN) in Basic Research of Central Auditory Processing: A Review," *Clinical Neurophysiology: Official Journal of the International Federation of Clinical Neurophysiology* 118, no. 12 (2007): 2544–2590, http://doi.org/10.1016/j.clinph.2007.04.026.

5. H. Honing, *Musical Cognition: A Science of Listening*, 2nd ed. (New Brunswick, NJ: Transaction Publishers, 2013), 87–97; originally published in Dutch as *Iedereen is muzikaal: Wat we weten over het luisteren naar muziek*, 2nd ed. (Amsterdam: Nieuw Amsterdam, 2012), 105–122.

6. I. Winkler, G. P. Háden, O. Ladinig, I. Sziller, and H. Honing, "Newborn Infants Detect the Beat in Music," *Proceedings of the National Academy of Sciences of the United States of America* 106, no. 7 (2009): 2468–2471, http://doi.org/10.1073/pnas.0809035106.

7. W. Zarco, H. Merchant, L. Prado, and J. C. Mendez, "Subsecond Timing in Primates: Comparison of Interval Production between Human Subjects and Rhesus Monkeys," *Journal of Neurophysiology* 102, no. 6 (2009): 3191–3202, http://doi.org/10.1152/jn.00066.2009.

8. W. T. Fitch, "The Biology and Evolution of Rhythm: Unravelling a Paradox," in *Language and Music as Cognitive Systems*, ed. P. Rebuschat, M. A. Rohrmeier, J. A. Hawkins, and I. Cross (Oxford: Oxford University Press, 2012), 73–95.

9. M. A. Arbib, ed., *Language, Music, and the Brain: A Mysterious Relationship* (Cambridge, MA: MIT Press, 2013).

10. A. Ueno, S. Hirata, K. Fuwa, K. Sugama, K. Kusunoki, G. Matsuda, H. Fukushima, K. Hiraki, M. Tomonaga, and T. Hasegawa, "Auditory ERPS to Stimulus Deviance in an Awake Chimpanzee (*Pan troglodytes*): Toward Hominid Cognitive Neurosciences," *PLOS One* 3, no. 1 (2008): 5.

11. H. Merchant, W. Zarco, O. Perez, L. Prado, and R. Bartolo, "Measuring Time with Different Neural Chronometers during a Synchronization-Continuation Task," *Proceedings of the National Academy of Sciences* 108, no. 49 (2011): 19784–19789, http://doi.org/10.1073/pnas.1112933108.

2 Mirroring

1. G. Rizzolatti and L. Craighero, "The Mirror-Neuron System," *Annual Review of Neuroscience* 27 (2004): 169–192, http://doi.org/10.1146/annurev.neuro.27.070203.144230.

2. D. Maestripieri, *Machiavellian Intelligence: How Rhesus Macaques and Humans Have Conquered the World* (Chicago: University of Chicago Press, 2007).

3 Beat Deaf

1. I. Peretz, "Neurobiology of Congenital Amusia," *Trends in Cognitive Sciences* 20, no. 11 (2016): 857–867, http://doi.org/10.1016/j.tics.2016.09.002.

2. J. Phillips-Silver, P. Toiviainen, N. Gosselin, O. Piché, S. Nozaradan, C. Palmer, and I. Peretz, "Born to Dance but Beat Deaf: A New Form of Congenital Amusia," *Neuropsychologia* 49, no. 5 (2011): 961–969, http://doi.org/10.1016/j.neuropsychologia.2011.02.002.

3. The interview can be found under the title "A Case of Congenital Beat Deafness?" at http://musiccognition.blogspot.com/2012/08/a-case-of-congenital-beat-deafness.html.

4. The Dutch-language VPRO documentary can be found under the title *De man zonder ritme* (The man without rhythm), at https://www.npo.nl/labyrint/14-12-2011/VPWON_1155438.

5. C. Palmer, P. Lidji, and I. Peretz, "Losing the Beat: Deficits in Temporal Coordination," *Philosophical Transactions of the Royal Society B: Biological Sciences* 369 (2014): 20130405, http://dx.doi.org/10.1098/rstb.2013.0405.

6. B. Mathias, P. Lidji, H. Honing, C. Palmer, and I. Peretz, "Electrical Brain Responses to Beat Irregularities in Two Cases of Beat Deafness," *Frontiers in Neuroscience* 10, no. 40 (2016): 1–13, http://doi.org/10.3389/fnins.2016.00040.

7. Peretz, "Neurobiology of Congenital Amusia," 857–867.

8. S. Dehaene and L. Cohen, "Cultural Recycling of Cortical Maps," *Neuron* 56, no. 2 (2007): 384–398, http://doi.org/10.1016/j.neuron.2007.10.004.

9. K. Tokarev, A. Tiunova, C. Scharff, and K. Anokhin, "Food for Song: Expression of c-Fos and ZENK in the Zebra Finch Song Nuclei during Food Aversion Learning," *PLOS One* 6, no. 6 (2011): e21157, http://doi.org/10.1371/journal.pone.0021157.

10. S. J. Gould and E. S. Vrba, "Exaptation: A Missing Term in the Science of Form," *Paleobiology* 8, no. 1 (1982): 4–15.

11. W. T. Fitch, "Musical Protolanguage: Darwin's Theory of Language Evolution Revisited," in *Birdsong, Speech, and Language: Exploring the Evolution of Mind and Brain*, ed. J. J. Bolhuis and M. Everaert (Cambridge, MA: MIT Press, 2013), 489–503.

4 Measuring the Beat

1. M. D. Hauser and P. Marler, "Food-Associated Calls in Rhesus Macaques (*Macaca mulatta*): I. Socioecological Factors," *Behavioral Ecology* 4, no. 3 (1993): 194–205.

2. P. F. Ferrari, E. Visalberghi, A. Paukner, L. Fogassi, A. Ruggiero, and S. J. Suomi, "Neonatal Imitation in Rhesus Macaques," *PLOS Biology* 4, no. 9 (2006): e302, http://doi.org/10.1371/journal.pbio.0040302.

3. J. C. Whitham, M. S. Gerald, and D. Maestripieri, "Intended Receivers and Functional Significance of Grunt and Girney Vocalizations in Free-Ranging Female Rhesus Macaques," *Ethology* 113, no. 9 (2007): 862–874, http://doi.org/10.1111/j.1439-0310.2007.01381.x.

4. J. H. McDermott and M. D. Hauser, "Nonhuman Primates Prefer Slow Tempos but Dislike Music Overall," *Cognition* 104, no. 3 (2007): 654–668, http://doi.org/10.1016/j.cognition.2006.07.011.

5. L. Robbins and S. W. Margulis, "The Effects of Auditory Enrichment on Gorillas," *Zoo Biology* 33, no. 3 (2014): 197–203, http://doi.org/10.1002/zoo.21127.

6. De Waal's article can be found on the internet using the search term "Chimps like listening to music," http://www.apa.org/news/press/releases/2014/06/chimps-music.aspx.

7. M. Mingle, T. Eppley, M. Campbell, K. Hall, V. Horner, and F. B. M. de Waal, "Chimpanzees Prefer African and Indian Music over Silence," *Journal of Experimental Psychology: Animal Learning and Cognition* 40, no. 4 (2014): 502–505, http://doi.org/10.1037/xan0000032.

8. The study ultimately appeared, with less-assertive claims, as E. W. Large and P. M. Gray, "Spontaneous Tempo and Rhythmic Entrainment in a Bonobo (*Pan paniscus*)," *Journal of Comparative Psychology* 129, no. 4 (2015): 317–328, http://dx.doi.org/10.1037/com0000011.

9. N. Konoike, A. Mikami, and S. Miyachi, "The Influence of Tempo upon the Rhythmic Motor Control in Macaque Monkeys," *Neuroscience Research* 74, no. 1 (2012): 4–7, http://doi.org/10.1016/j.neures.2012.06.002.

10. H. Honing, H. Merchant, G. P. Háden, L. Prado, and R. Bartolo, "Probing Beat Induction in Rhesus Monkeys: Is Beat Induction Species-Specific?" in *Proceedings of the 12th International Conference on Music Perception and Cognition*, ed. E. Cambouropoulos, C. Tsougras, P. Mavromatis, and K. Pastiadis (Thessaloniki: University of Thessaloniki, 2012), 454–455.

11. H. Honing, H. Merchant, G. P. Háden, L. Prado, and R. Bartolo, "Rhesus Monkeys (*Macaca mulatta*) Detect Rhythmic Groups in Music, but Not the Beat," *PLOS One* 7, no. 12 (2012): 1–10, http://doi.org/10.1371/journal.pone.0051369.

12. W. T. Fitch, B. de Boer, N. Mathur, and A. A. Ghazanfar, "Monkey Vocal Tracts Are Speech-Ready," *Science Advances* 2, no. 12 (2016): e1600723, http://doi.org/10.1126/sciadv.1600723.

13. J. A. Grahn and M. Brett, "Rhythm and Beat Perception in Motor Areas of the Brain," *Journal of Cognitive Neuroscience* 19, no. 5 (2007): 893–906, http://doi.org/10.1162/jocn.2007.19.5.893.

14. H. Merchant and H. Honing, "Are Non-human Primates Capable of Rhythmic Entrainment? Evidence for the Gradual Audiomotor Evolution Hypothesis," *Frontiers in Neuroscience* 7, no. 274 (2014): 1–8, http://doi.org/10.3389/fnins.2013.00274.

15. Y. Nagasaka, Z. C. Chao, N. Hasegawa, T. Notoya, and N. Fujii, "Spontaneous Synchronization of Arm Motion between Japanese Macaques," *Scientific Reports* 3, no. 1151 (2013), http://doi.org/10.1038/srep01151.

5 Ai and Ayumu

1. S. Inoue and T. Matsuzawa, "Working Memory of Numerals in Chimpanzees," *Current Biology* 17, no. 23 (2007): 1004–1005, http://doi.org/10.1016/j.cub.2007.10.027.

2. B. Merker, G. S. Madison, and P. Eckerdal, "On the Role and Origin of Isochrony in Human Rhythmic Entrainment," *Cortex: A Journal Devoted to the Study of the Nervous System and Behavior* 45, no. 1 (2009): 4–17, http://doi.org/10.1016/j.cortex.2008.06.011.

3. Y. Hattori, M. Tomonaga, and T. Matsuzawa, "Spontaneous Synchronized Tapping to an Auditory Rhythm in a Chimpanzee," *Scientific Reports* 3 (2013): 1566, http://doi.org/10.1038/srep01566.

4. Y. Hattori, M. Tomonaga, and T. Matsuzawa, "Distractor Effect of Auditory Rhythms on Self-Paced Tapping in Chimpanzees and Humans," *PLOS One* 10, no. 7 (2015): 1–17, https://doi.org/10.1371/journal.pone.0130682.

6 Supernormal Stimulus

1. J. J. Bolhuis and M. Everaert, eds., *Birdsong, Speech, and Language* (Cambridge, MA: MIT Press, 2013).

2. A. D. Patel, "Musical Rhythm, Linguistic Rhythm, and Human Evolution," *Music Perception* 24, no. 1 (2006): 99–104, http://doi.org/10.1525/mp.2006.24.1.99.

3. Species hierarchy after E. D. Jarvis, "Selection for and against Vocal Learning in Birds and Mammals," *Ornithological Science* 5, no. 1 (2006): 5–14, fig. 1, https://doi.org/10.2326/osj.5.5.

4. N. Tinbergen and A. Perdeck, "On the Stimulus Situation Releasing the Begging Response in the Newly Hatched Herring Gull Chick (*Larus argentatus argentatus* Pont)," *Behavior* 3, no. 1 (1950): 1–39.

5. N. Tinbergen, *The Study of Instinct* (Oxford: Clarendon Press, 1951); *The Herring Gull's World* (London: Collins, 1953).

6. C. ten Cate, W. S. Bruins, J. den Ouden, T. Egberts, H. Neevel, M. Spierings, K. van der Burg, and A. G. Brokerhof, "Tinbergen Revisited: A Replication and Extension of Experiments on the Beak Color Preferences of Herring Gull Chicks," *Animal Behavior* 77, no. 4 (2009): 795–802.

7. F. L. Bouwer, T. L van Zuijen, and H. Honing, "Beat Processing Is Pre-attentive for Metrically Simple Rhythms with Clear Accents: An ERP Study," *PLOS One* 9, no. 5 (2014): e97467, http://doi.org/10.1371/journal.pone.0097467.

8. H. Honing, *Musical Cognition: A Science of Listening*, 2nd ed. (New Brunswick, NJ: Transaction Publishers, 2013), 66; published in Dutch as *Iedereen is muzikaal: Wat we weten over het luisteren naar muziek*, 2nd ed. (Amsterdam: Nieuw Amsterdam, 2012), 85.

9. E. J. Wagenmakers, "A Practical Solution to the Pervasive Problems of *p* Values," *Psychonomic Bulletin & Review 14*, no. 5: (2007): 779–804, http://doi.org/10.3758/BF03194105.

7 Snowball

1. The video clip was posted on YouTube on October 15, 2007, and has since attracted millions of viewers. It can be found using the search terms "Snowball, dancing, cockatoo."

2. J. R. Lucas, T. M. Freeberg, G. R. Long, and A. Krishnan, "Seasonal Variation in Avian Auditory Evoked Responses to Tones: A Comparative Analysis of Carolina Chickadees, Tufted Titmice, and White-Breasted Nuthatches," *Journal of Comparative Physiology A: Neuroethology, Sensory, Neural, and Behavioral Physiology* 193, no. 2 (2007): 201–215, http://doi.org/10.1007/s00359-006-0180-z.

3. E. Vallet and M. Kreutzer, "Female Canaries Are Sexually Responsive to Special Song Phrases," *Animal Behaviour* 49, no. 6 (1995): 1603–1610, http://doi.org/10.1016/0003-3472(95)90082-9.

4. R. J. Zatorre and V. N. Salimpoor, "From Perception to Pleasure: Music and Its Neural Substrates," *Proceedings of the National Academy of Sciences of the United States of America* 110, supplement 2 (2013): 10430–10437, http://doi.org/10.1073/pnas.1301228110.

5. B. A. Alward, M. L. Rouse, J. Balthazart, and G. F. Ball, "Testosterone Regulates Birdsong in an Anatomically Specific Manner," *Animal Behaviour* 124 (2017): 291–298, http://doi.org/10.1016/j.anbehav.2016.09.013.

6. M. Hultcrantz, R. Simonoska, and A. E. Stenberg, "Estrogen and Hearing: A Summary of Recent Investigations," *Acta Oto-Laryngologica* 126, no. 1 (2006): 10–14, http://doi.org/10.1080/00016480510038617.

7. H. Richner, "Interval Singing Links to Phenotypic Quality in a Songbird," *Proceedings of the National Academy of Sciences of the United States of America* 113, no. 45 (2016): 12763–12767, http://doi.org/10.1073/pnas.1610062113. See also the chapter "Perfect Pitch" later in this book (in the section "To Listen like a Songbird").

8. M. A. Bellis, T. Hennell, C. Lushey, K. Hughes, K. Tocque, and J. R. Ashton, "Elvis to Eminem: Quantifying the Price of Fame through Early Mortality of European and North American Rock and Pop Stars," *Journal of Epidemiology and Community Health* 61 (2007): 896–901, http://doi.org/10.1136/jech.2007.059915.

9. B. D. Charlton, P. Filippi, and W. T. Fitch, "Do Women Prefer More Complex Music around Ovulation?" *PLOS One* 7, no. 4 (2012): e35626, http://doi.org/10.1371/journal.pone.0035626.

10. B. D. Charlton, "Menstrual Cycle Phase Alters Women's Sexual Preferences for Composers of More Complex Music," *Proceedings of the Royal Society B: Biological Sciences* 281, no. 1784 (2014), http://doi.org/10.1098/rspb.2014.0403.

11. M. A. Mosing, K. J. H. Verweij, G. Madison, N. L. Pedersen, B. P. Zietsch, and F. Ullén, "Did Sexual Selection Shape Human Music? Testing Predictions from the Sexual Selection Hypothesis of Music Evolution Using a Large Genetically Informative Sample of Over 10,000 Twins," *Evolution and Human Behavior* 36, no. 5 (2015): 359–366, http://doi.org/10.1016/j.evolhumbehav.2015.02.004.

12. A. D. Patel, J. R. Iversen, M. R. Bregman, and I. Schulz, "Experimental Evidence for Synchronization to a Musical Beat in a Nonhuman Animal," *Current Biology* 19, no. 10 (2009): 827–830, http://doi.org/10.1016/j.cub.2009.03.038.

13. A. Schachner, T. F. Brady, I. M. Pepperberg, and M. D. Hauser, "Spontaneous Motor Entrainment to Music in Multiple Vocal Mimicking Species," *Current Biology* 19, no. 10 (2009): 831–836, http://doi.org/10.1016/j.cub.2009.03.061.

14. W. T. Fitch, "Biology of Music: Another One Bites the Dust," *Current Biology* 19, no. 10 (2009): R403–R404, http://doi.org/10.1016/j.cub.2009.04.004.

15. A. Hasegawa, K. Okanoya, T. Hasegawa, and Y. Seki, "Rhythmic Synchronization Tapping to an Audio-Visual Metronome in Budgerigars," *Scientific Reports* 1 (2011): 1–8, http://doi.org/10.1038/srep00120.

8 The Dialect of Song

1. P. Marler and H. Slabbekoorn, eds., *Nature's Music: The Science of Birdsong* (London: Academic Press, 2004).

2. P. Marler, "Science and Birdsong: The Good Old Days," in *Nature's Music: The Science of Birdsong*, ed. P. Marler and H. Slabbekoorn (London: Academic Press, 2004), 1–38.

3. C. ten Cate, "Assessing the Uniqueness of Language: Animal Grammatical Abilities Take Center Stage," *Psychonomic Bulletin and Review* 24 (2017): 91–96, http://doi.org/10.3758/s13423-016-1091-9.

4. H. Honing, "Muzikaliteit gaat aan muziek én taal vooraf," *Blind* 43 (2016): 1–8.

5. J. J. Bolhuis and M. Everaert, eds., *Birdsong, Speech, and Language* (Cambridge, MA: MIT Press, 2013).

6. S. E. Trehub, "The Developmental Origins of Musicality," *Nature Neuroscience* 6 (2003): 669–673.

7. S. L. Mattys, P. W. Jusczyk, P. A. Luce, and J. L. Morgan, "Phonotactic and Prosodic Effects on Word Segmentation in Infants," *Cognitive Psychology* 38, no. 4 (1999): 465–494.

8. S. Pinker, *How the Mind Works* (New York: Norton, 1997).

9. R. J. Dooling, "Audition: Can Birds Hear Everything They Sing?" in *Nature's Music: The Science of Birdsong*, ed. P. Marler and H. Slabbekoorn (London: Academic Press, 2004), 206–225, fig. 7.2.

10. G. P. Háden, G. Stefanics, M. D. Vestergaard, S. L. Denham, I. Sziller, and I. Winkler, "Timbre-Independent Extraction of Pitch in Newborn Infants," *Psychophysiology* 46, no. 1 (2009): 69–74, http://doi.org/10.1111/j.1469-8986.2008.00749.x.

11. R. Weisman, M. Hoeschele, and C. B. Sturdy, "A Comparative Analysis of Auditory Perception in Humans and Songbirds: A Modular Approach," *Behavioral Processes* 104 (2014): 35–43, http://doi.org/10.1016/j.beproc.2014.02.006.

12. J. Nicolai, C. Gundacker, K. Teeselink, and H. R. Güttinger, "Human Melody Singing by Bullfinches (*Pyrrhula pyrrula*) Gives Hints about a Cognitive Note Sequence Processing," *Animal Cognition* 17, no. 1 (2014): 143–155, http://doi.org/10.1007/s10071-013-0647-6.

9 Perfect Pitch

1. The video can be found on the internet using the search terms "dog, perfect, pitch."

2. D. J. Levitin and S. E. Rogers, "Absolute Pitch: Perception, Coding, and Controversies," *Trends in Cognitive Sciences* 9, no. 1 (2005): 26–33, http://doi.org/10.1016/j.tics.2004.11.007.

3. B. Gingras, H. Honing, I. Peretz, L. J. Trainor, and S. E. Fisher, "Defining the Biological Bases of Individual Differences in Musicality," *Philosophical Transactions of the Royal Society B: Biological Sciences* 370, no. 1664 (2015), http://doi.org/10.1098/rstb.2014.0092.

4. R. G. Weisman, D. Mewhort, M. Hoeschele, and C. B. Sturdy, "New Perspectives on Absolute Pitch in Birds and Mammals," in *The Oxford Handbook of Comparative*

Cognition, 2nd ed., ed. E. A. Wasserman and T. R. Zentall (Oxford: Oxford University Press, 2012), 67–79.

5. A. L.-C. Wang, "The Shazam Music Recognition Service," *Communications of the ACM* 49 (2006): 44–48, http://doi.org/10.1145/1145287.1145312.

6. S. H. Hulse, J. Cynx, and J. Humpal, "Absolute and Relative Pitch Discrimination in Serial Pitch Perception by Birds," *Journal of Experimental Psychology: General* 113, no. 1 (1984): 38–54, http://doi.org/10.1037/0096-3445.113.1.38.

7. A. D. Patel, "Why Doesn't a Songbird (the European Starling) Use Pitch to Recognize Tone Sequences? The Informational Independence Hypothesis," *Comparative Cognition and Behavior Reviews* 12 (2017): 19–32, http://doi.org/10.3819/CCBR.2017.120003.

8. M. R. Bregman, A. D. Patel, and T. Q. Gentner, "Songbirds Use Spectral Shape, Not Pitch, for Sound Pattern Recognition," *Proceedings of the National Academy of Sciences* 113, no. 6 (2016): 1666–1671, http://doi.org/10.1073/pnas.1515380113.

9. J. E. Elie and F. E. Theunissen, "The Vocal Repertoire of the Domesticated Zebra Finch: A Data-Driven Approach to Decipher the Information-Bearing Acoustic Features of Communication Signals," *Animal Cognition* 19, no. 2 (2016): 285–315, http://doi.org/10.1007/s10071-015-0933-6.

10. R. V. Shannon, "Is Birdsong More like Speech or Music?" *Trends in Cognitive Sciences* 20, no. 4 (2016): 245–247, http://doi.org/10.1016/j.tics.2016.02.004.

11. M. J. Spierings and C. ten Cate, "Zebra Finches Are Sensitive to Prosodic Features of Human Speech," *Proceedings of the Royal Society B: Biological Sciences* 281, no. 1787 (2014): 1–7, http://doi.org/10.1098/rspb.2014.0480.

12. J. Kursell, "Experiments on Tone Color in Music and Acoustics: Helmholtz, Schoenberg, and *Klangfarbenmelodie*," *Osiris* 28, no. 1 (2013): 191–211, http://doi.org/10.1086/671377.

13. J. H. McDermott and A. J. Oxenham, "Music Perception, Pitch, and the Auditory System," *Current Opinion in Neurobiology* 18, no. 4 (2008): 452–463, http://doi.org/10.1016/j.conb.2008.09.005.

14. K. I. Nagel, H. M. McLendon, and A. J. Doupe, "Differential Influence of Frequency, Timing, and Intensity Cues in a Complex Acoustic Categorization Task," *Journal of Neurophysiology* 104, no. 3 (2010): 1426–1437, http://doi.org/10.1152/jn.00028.2010.

15. J. van der Aa, H. Honing, and C. ten Cate, "The Perception of Regularity in an Isochronous Stimulus in Zebra Finches (*Taeniopygia guttata*) and Humans," *Behavioral Processes* 115 (2015): 37–45, http://doi.org/10.1016/j.beproc.2015.02.018.

16. J. J. Gibson, "Events Are Perceivable but Time Is Not," in *The Study of Time II*, ed. J. T. Fraser and N. Lawrence (New York: Springer, 1975), 295–301, http://doi.org/10.1007/978-3-642-50121-0_22.

17. J. Sivic, B. Russell, and A. Efros, *Discovering Object Categories in Image Collections*, Computer Science and Artificial Intelligence Laboratory Technical Report (Cambridge, MA: Massachusetts Institute of Technology, 2005).

18. K. Shinozuka, H. Ono, and S. Watanabe, "Reinforcing and Discriminative Stimulus Properties of Music in Goldfish," *Behavioral Processes* 99 (2013): 26–33, http://doi.org/10.1016/j.beproc.2013.06.009.

19. A. R. Chase, "Music Discriminations by Carp (*Cyprinus carpio*)," *Animal Learning and Behavior* 29, no. 4 (2001): 336–353.

20. D. Porter and A. Neuringer, "Music Discriminations by Pigeons," *Journal of Experimental Psychology: Animal Behavior Processes* 10, no. 2 (1984): 138–148.

10 Rio and Ronan

1. C. Reichmuth Kastak and R. J. Schusterman, "Long-Term Memory for Concepts in a California Sea Lion (*Zalophus californianus*)," *Animal Cognition* 5, no. 4 (2002): 225–232, http://doi.org/10.1007/s10071-002-0153-8.

2. R. J. Schusterman, "Vocal Learning in Mammals with Special Emphasis on Pinnipeds," in *The Evolution of Communicative Flexibility: Complexity, Creativity, and Adaptability in Human and Animal Communication*, ed. D. K. Oller and U. Gribel (Cambridge, MA: MIT Press, 2008), 41–70.

3. P. F. Cook, A. Rouse, M. Wilson, and C. Reichmuth, "A California Sea Lion (*Zalophus californianus*) Can Keep the Beat: Motor Entrainment to Rhythmic Auditory Stimuli in a Non Vocal Mimic," *Journal of Comparative Psychology* 127, no. 4 (2013): 412–427, http://doi.org/10.1037/a0032345.

4. R. J. Schusterman and R. F. Balliet, "Conditioned Vocalizations as a Technique for Determining Visual Acuity Thresholds in Sea Lions," *Science* 169, no. 3944 (1970): 498–501, http://doi.org/10.1126/science.169.3944.498.

5. U. Arnason, A. Gullberg, A. Janke, M. Kullberg, N. Lehman, E. A. Petrov, and R. Väinölä, "Pinniped Phylogeny and a New Hypothesis for Their Origin and Dispersal," *Molecular Phylogenetics and Evolution* 41, no. 2 (2006): 345–354, http://doi.org/10.1016/j.ympev.2006.05.022.

6. Relative likelihood: $1/5 \times (25{,}000 - 7{,}100) / 7{,}100 \approx 0.5$.

7. A. Ravignani and P. F. Cook, "The Evolutionary Biology of Dance without Frills," *Current Biology* 26, no. 19 (2016): R878–R879, https://doi.org/10.1016/j.cub.2016.07.076.

8. C. I. Petkov and E. D. Jarvis, "Birds, Primates, and Spoken Language Origins: Behavioral Phenotypes and Neurobiological Substrates," *Frontiers in Evolutionary Neuroscience* 4 (2012): 12, http://doi.org/10.3389/fnevo.2012.00012.

9. A. D. Patel and J. R. Iversen, "The Evolutionary Neuroscience of Musical Beat Perception: The Action Simulation for Auditory Prediction (ASAP) Hypothesis," *Frontiers in Systems Neuroscience* 8 (2014): 57, http://doi.org/10.3389/fnsys.2014.00057.

10. H. Honing and H. Merchant, "Differences in Auditory Timing between Human and Non-human Primates," *Behavioral and Brain Sciences* 27, no. 6 (2014): 557–558.

Afterword

1. B. H. Repp, "Some Cognitive and Perceptual Aspects of Speech and Music," in *Music, Language, Speech, and Brain*, ed. J. Sundberg, L. Nord, and R. Carlson (Cambridge: Macmillan, 1991), 257–268.

2. R. J. Zatorre and V. N. Salimpoor, "From Perception to Pleasure: Music and Its Neural Substrates," *Proceedings of the National Academy of Sciences of the United States of America* 110, supplement 2 (2013): 10430–10437, http://doi.org/10.1073/pnas.1301228110.

3. A. D. Patel, "Music as a Transformative Technology of the Mind: An Update," in *The Origins of Musicality*, ed. H. Honing (Cambridge, MA: MIT Press, 2018), 113–126.

4. J. Goudsblom, *Fire and Civilization* (London: Penguin, 1995).

5. B. Gingras, H. Honing, I. Peretz, L. J. Trainor, and S. E. Fisher, "Defining the Biological Bases of Individual Differences in Musicality," *Philosophical Transactions of the Royal Society B: Biological Sciences* 370, no. 1664 (2015), http://doi.org/10.1098/rstb.2014.0092.

6. H. Honing, "Musicality as an Upbeat to Music: Introduction and Research Agenda," in *The Origins of Musicality*, ed. H. Honing (Cambridge, MA: MIT Press, 2018), 3–20.

7. The special issue on musicality appeared as H. Honing, C. ten Cate, I. Peretz, and S. E. Trehub, "Without It No Music: Cognition, Biology, and Evolution of Musicality," *Philosophical Transactions of the Royal Society B: Biological Sciences* 370, no. 1664 (2015), http://doi.org/10.1098/rstb.2014.0088. An edited and expanded version appeared as H. Honing, ed., *The Origins of Musicality* (Cambridge, MA: MIT Press, 2018).

Explanatory Note

1. An annual Dutch lecture given by a leading scientist, designer, or writer on a subject at the interface of society, health care, and culture. The series is named after Pieter van Foreest, the so-called Dutch Hippocrates, a prominent sixteenth-century Dutch physician.

References

Alward, B. A., M. L. Rouse, J. Balthazart, and G. F. Ball. 2017. Testosterone regulates birdsong in an anatomically specific manner. *Animal Behaviour* 124:291–298. doi:10.1016/j.anbehav.2016.09.013.

Arbib, M. A., ed. 2013. *Language, Music, and the Brain: A Mysterious Relationship*. Cambridge, MA: MIT Press.

Arnason, U., A. Gullberg, A. Janke, M. Kullberg, N. Lehman, E. A. Petrov, and R. Väinölä. 2006. Pinniped phylogeny and a new hypothesis for their origin and dispersal. *Molecular Phylogenetics and Evolution* 41 (2): 345–354. doi:10.1016/j.ympev.2006.05.022.

Bellis, M. A., T. Hennell, C. Lushey, K. Hughes, K. Tocque, and J. R. Ashton. 2007. Elvis to Eminem: Quantifying the price of fame through early mortality of European and North American rock and pop stars. *Journal of Epidemiology and Community Health* 61:896–901. doi:10.1136/jech.2007.059915.

Bolhuis, J. J., and M. Everaert, eds. 2013. *Birdsong, Speech, and Language*. Cambridge, MA: MIT Press.

Bouwer, F. L., T. L. van Zuijen, and H. Honing. 2014. Beat processing is pre-attentive for metrically simple rhythms with clear accents: An ERP study. *PLOS One* 9 (5): e97467. doi:10.1371/journal.pone.0097467.

Bregman, M. R., A. D. Patel, and T. Q. Gentner. 2016. Songbirds use spectral shape, not pitch, for sound pattern recognition. *Proceedings of the National Academy of Sciences of the United States of America* 113 (6): 1666–1671. doi:10.1073/pnas.1515380113.

Charlton, B. D. 2014. Menstrual cycle phase alters women's sexual preferences for composers of more complex music. *Proceedings of the Royal Society B: Biological Sciences* 281 (1784): 20140403. doi:10.1098/rspb.2014.0403.

Charlton, B. D., P. Filippi, and W. T. Fitch. 2012. Do women prefer more complex music around ovulation? *PLOS One* 7 (4): e35626. doi:10.1371/journal.pone.0035626.

Chase, A. R. 2001. Music discriminations by carp (*Cyprinus carpio*). *Animal Learning and Behavior* 29 (4): 336–353.

Cook, P. F., A. Rouse, M. Wilson, and C. Reichmuth. 2013. A California sea lion (*Zalophus californianus*) can keep the beat: Motor entrainment to rhythmic auditory stimuli in a non vocal mimic. *Journal of Comparative Psychology* 127 (2): 412–427. doi:10.1037/a0032345.

Darwin, C. 1871. *The Descent of Man, and Selection in Relation to Sex*. London: John Murray.

Dehaene, S., and L. Cohen. 2007. Cultural recycling of cortical maps. *Neuron* 56 (2): 384–398. doi:10.1016/j.neuron.2007.10.004.

Dooling, R. J. 2004. Audition: Can birds hear everything they sing? In *Nature's Music: The Science of Birdsong*, ed. P. Marler and H. Slabbekoorn, 206–225. London: Elsevier Academic Press.

Elie, J. E., and F. E. Theunissen. 2016. The vocal repertoire of the domesticated zebra finch: A data-driven approach to decipher the information-bearing acoustic features of communication signals. *Animal Cognition* 19 (2): 285–315. doi:10.1007/s10071-015-0933-6.

Ferrari, P. F., E. Visalberghi, A. Paukner, L. Fogassi, A. Ruggiero, and S. J. Suomi. 2006. Neonatal imitation in rhesus macaques. *PLOS Biology* 4 (9): e302. doi:10.1371/journal.pbio.0040302.

Fitch, W. T. 2009. Biology of music: Another one bites the dust. *Current Biology* 19 (10): R403–R404. doi:10.1016/j.cub.2009.04.004.

Fitch, W. T. 2012. The biology and evolution of rhythm: Unravelling a paradox. In *Language and Music as Cognitive Systems*, ed. P. Rebuschat, M. A. Rohrmeier, J. A. Hawkins, and I. Cross, 73–95. Oxford: Oxford University Press.

Fitch, W. T. 2013. Musical protolanguage: Darwin's theory of language evolution revisited. In *Birdsong, Speech, and Language: Exploring the Evolution of Mind and Brain*, ed. J. J. Bolhuis and M. Everaert, 489–503. Cambridge, MA: MIT Press.

Fitch, W. T., B. de Boer, N. Mathur, and A. A. Ghazanfar. 2016. Monkey vocal tracts are speech-ready. *Science Advances* 2 (12): e1600723. doi:10.1126/sciadv.1600723.

Gibson, J. J. 1975. Events are perceivable but time is not. In *The Study of Time II*, ed. J. T. Fraser and N. Lawrence, 295–301. New York: Springer. doi:10.1007/978-3-642-50121-0_22.

Gingras, B., H. Honing, I. Peretz, L. J. Trainor, and S. E. Fisher. 2015. Defining the biological bases of individual differences in musicality. *Philosophical Transactions of the Royal Society B: Biological Sciences* 370 (1664): 20140092. doi:10.1098/rstb.2014.0092.

Goudsblom, J. 1995. *Fire and Civilization*. London: Penguin Books.

Gould, S. J., and E. S. Vrba. 1982. Exaptation: A missing term in the science of form. *Paleobiology* 8 (1): 4–15.

Grahn, J. A., and M. Brett. 2007. Rhythm and beat perception in motor areas of the brain. *Journal of Cognitive Neuroscience* 19 (5): 893–906. doi:10.1162/jocn.2007.19.5.893.

Háden, G. P., G. Stefanics, M. D. Vestergaard, S. L. Denham, I. Sziller, and I. Winkler. 2009. Timbre-independent extraction of pitch in newborn infants. *Psychophysiology* 46 (1): 69–74. doi:10.1111/j.1469-8986.2008.00749.x.

Hasegawa, A., K. Okanoya, T. Hasegawa, and Y. Seki. 2011. Rhythmic synchronization tapping to an audio-visual metronome in budgerigars. *Scientific Reports* 1:1–8. doi:10.1038/srep00120.

Hattori, Y., M. Tomonaga, and T. Matsuzawa. 2013. Spontaneous synchronized tapping to an auditory rhythm in a chimpanzee. *Scientific Reports* 3:1566. doi:10.1038/srep01566.

Hattori, Y., M. Tomonaga, and T. Matsuzawa. 2015. Distractor effect of auditory rhythms on self-paced tapping in chimpanzees and humans. *PLOS One* 10 (7): 1–17. https://doi.org/10.1371/journal.pone.0130682.

Hauser, M. D., and P. Marler. 1993. Food-associated calls in rhesus macaques (*Macaca mulatta*): I. Socioecological factors. *Behavioural Ecology* 4 (3): 194–205.

Honing, H. 2012. Without it no music: Beat induction as a fundamental musical trait. *Annals of the New York Academy of Sciences* 1252 (1): 85–91. doi:10.1111/j.1749-6632.2011.06402.x.

Honing, H. 2013. *Musical Cognition: A Science of Listening*. 2nd ed. New Brunswick, NJ: Transaction Publishers. Published in Dutch as *Iedereen is muzikaal: Wat we weten over het luisteren naar muziek*, 2nd ed. (Amsterdam: Nieuw Amsterdam, 2012).

Honing, H. 2016. Muzikaliteit gaat aan muziek én taal vooraf. *Blind* 43:1–8.

Honing, H. 2018. Musicality as an upbeat to music: Introduction and research agenda. In *The Origins of Musicality*, ed. H. Honing, 3–20. Cambridge, MA: MIT Press.

Honing, H., ed. 2018. *The Origins of Musicality*. Cambridge, MA: MIT Press.

Honing, H., and H. Merchant. 2014. Differences in auditory timing between human and non-human primates. *Behavioral and Brain Sciences* 27 (6): 557–558.

Honing, H., H. Merchant, G. P. Háden, L. Prado, and R. Bartolo. 2012a. Probing beat induction in rhesus monkeys: Is beat induction species-specific? *Proceedings of the 12th International Conference of Music Perception and Cognition*, ed. E. Cambouropoulos,

C. Tsougras, P. Mavromatis, and K. Pastiadis, 454–455. Thessaloniki: University of Thessaloniki.

Honing, H., H. Merchant, G. P. Háden, L. Prado, and R. Bartolo. 2012b. Rhesus monkeys (*Macaca mulatta*) detect rhythmic groups in music, but not the beat. *PLOS One* 7 (12): 1–10. doi:10.1371/journal.pone.0051369.

Honing, H., C. ten Cate, I. Peretz, and S. E. Trehub. 2015. Without it no music: Cognition, biology, and evolution of musicality. *Philosophical Transactions of the Royal Society B: Biological Sciences* 370 (1664): 20140088. Doi:10.1098/rstb.2014.0088.

Hulse, S. H., J. Cynx, and J. Humpal. 1984. Absolute and relative pitch discrimination in serial pitch perception by birds. *Journal of Experimental Psychology: General* 113 (1): 38–54. doi:10.1037/0096-3445.113.1.38.

Hultcrantz, M., R. Simonoska, and A. E. Stenberg. 2006. Estrogen and hearing: A summary of recent investigations. *Acta Oto-Laryngologica* 126 (1): 10–14. doi:10.1080/00016480510038617.

Inoue, S., and T. Matsuzawa. 2007. Working memory of numerals in chimpanzees. *Current Biology* 17 (23): 1004–1005. doi:10.1016/j.cub.2007.10.027.

Jarvis, E. D. 2006. Selection for and against vocal learning in birds and mammals. *Ornithological Science* 5 (1): 5–14. https://doi.org/10.2326/osj.5.5.

Konoike, N., A. Mikami, and S. Miyachi. 2012. The influence of tempo upon the rhythmic motor control in macaque monkeys. *Neuroscience Research* 74 (1): 4–7. doi:10.1016/j.neures.2012.06.002.

Kursell, J. 2013. Experiments on tone color in music and acoustics: Helmholtz, Schoenberg, and *Klangfarbenmelodie*. *Osiris* 28 (1): 191–211. doi:10.1086/671377.

Large, E. W., and P. M. Gray. 2015. Spontaneous tempo and rhythmic entrainment in a bonobo (*Pan paniscus*). *Journal of Comparative Psychology* 129 (4): 317–328. doi:10.1037/com0000011.

Levitin, D. J., and S. E. Rogers. 2005. Absolute pitch: Perception, coding, and controversies. *Trends in Cognitive Sciences* 9 (1): 26–33. doi:10.1016/j.tics.2004.11.007.

Lomax, A., and N. Berkowitz. 1972. The evolutionary taxonomy of culture. *Science* 177 (4045): 228–239. doi:10.1126/science.177.4045.228.

Lucas, J. R., T. M. Freeberg, G. R. Long, and A. Krishnan. 2007. Seasonal variation in avian auditory evoked responses to tones: A comparative analysis of Carolina chickadees, tufted titmice, and white-breasted nuthatches. *Journal of Comparative Physiology A: Neuroethology, Sensory, Neural, and Behavioral Physiology* 193 (2): 201–215. doi:10.1007/s00359-006-0180-z.

Maestripieri, D. 2007. *Machiavellian Intelligence: How Rhesus Macaques and Humans Have Conquered the World*. Chicago: University of Chicago Press.

Marler, P. 2004. Science and birdsong: The good old days. In *Nature's Music: The Science of Birdsong*, ed. P. Marler and H. Slabbekoorn, 1–38. London: Academic Press.

Marler, P., and H. Slabbekoorn, eds. 2004. *Nature's Music: The Science of Birdsong*. London: Academic Press.

Mathias, B., P. Lidji, H. Honing, C. Palmer, and I. Peretz. 2016. Electrical brain responses to beat irregularities in two cases of beat deafness. *Frontiers in Neuroscience* 10 (40): 1–13. doi:10.3389/fnins.2016.00040.

Mattys, S. L., P. W. Jusczyk, P. A. Luce, and J. L. Morgan. 1999. Phonotactic and prosodic effects on word segmentation in infants. *Cognitive Psychology* 38 (4): 465–494.

McDermott, J. H., and M. D. Hauser. 2007. Nonhuman primates prefer slow tempos but dislike music overall. *Cognition* 104 (3): 654–668. doi:10.1016/j.cognition.2006.07.011.

McDermott, J. H., and A. J. Oxenham. 2008. Music perception, pitch, and the auditory system. *Current Opinion in Neurobiology* 18 (4): 452–463. doi:10.1016/j.conb.2008.09.005.

Merchant, H., and H. Honing. 2014. Are non-human primates capable of rhythmic entrainment? Evidence for the gradual audiomotor evolution hypothesis. *Frontiers in Neuroscience* 7 (274): 1–8. doi:/10.3389/fnins.2013.00274.

Merchant, H., W. Zarco, O. Perez, L. Prado, and R. Bartolo. 2011. Measuring time with different neural chronometers during a synchronization-continuation task. *Proceedings of the National Academy of Sciences of the United States of America* 108 (49): 19784–19789. doi:10.1073/pnas.1112933108.

Merker, B., G. S. Madison, and P. Eckerdal. 2009. On the role and origin of isochrony in human rhythmic entrainment. *Cortex* 45 (1): 4–17. doi:10.1016/j.cortex.2008.06.011.

Mingle, M., T. Eppley, M. Campbell, K. Hall, V. Horner, and F. B. M. de Waal. 2014. Chimpanzees prefer African and Indian music over silence. *Journal of Experimental Psychology: Animal Learning and Cognition* 40 (4): 502–505. doi:10.1037/xan0000032.

Mosing, M. A., K. J. H. H. Verweij, G. Madison, N. L. Pedersen, B. P. Zietsch, and F. Ullén. 2015. Did sexual selection shape human music? Testing predictions from the sexual selection hypothesis of music evolution using a large genetically informative sample of over 10,000 twins. *Evolution and Human Behavior* 36 (5): 359–366. doi:10.1016/j.evolhumbehav.2015.02.004.

Näätänen, R., P. Paavilainen, T. Rinne, and K. Alho. 2007. The mismatch negativity (MMN) in basic research of central auditory processing: A review. *Clinical Neurophysiology: Official Journal of the International Federation of Clinical Neurophysiology* 118 (12): 2544–2590. doi:10.1016/j.clinph.2007.04.026.

Nagasaka, Y., Z. C. Chao, N. Hasegawa, T. Notoya, and N. Fujii. 2013. Spontaneous synchronization of arm motion between Japanese macaques. *Scientific Reports* 3:1151. doi:10.1038/srep01151.

Nagel, K. I. K., H. M. McLendon, and A. J. Doupe. 2010. Differential influence of frequency, timing, and intensity cues in a complex acoustic categorization task. *Journal of Neurophysiology* 104 (3): 1426–1437. doi:10.1152/jn.00028.2010.

Nettl, B. 2000. An ethnomusicologist contemplates universals in musical sound and musical culture. In *The Origins of Music*, ed. N. L. Wallin, B. Merker, and S. Brown, 463–472. Cambridge, MA: MIT Press.

Nicolai, J., C. Gundacker, K. Teeselink, and H. R. Güttinger. 2014. Human melody singing by bullfinches (*Pyrrhula pyrrula*) gives hints about a cognitive note sequence processing. *Animal Cognition* 17 (1): 143–155. doi:10.1007/s10071-013-0647-6.

Palmer, C., P. Lidji, and I. Peretz. 2014. Losing the beat: Deficits in temporal coordination. *Philosophical Transactions of the Royal Society B: Biological Sciences* 369:20130405. doi:10.1098/rstb.2013.0405.

Patel, A. D. 2006. Musical rhythm, linguistic rhythm, and human evolution. *Music Perception* 24 (1): 99–104. doi:10.1525/mp.2006.24.1.99.

Patel, A. D. 2017. Why doesn't a songbird (the European starling) use pitch to recognize tone sequences? The informational independence hypothesis. *Comparative Cognition and Behavior Reviews* 12:19–32. doi:10.3819/CCBR.2017.120003.

Patel, A. D. 2018. Music as a transformative technology of the mind: An update. In *The Origins of Musicality*, ed. H. Honing, 113–126. Cambridge, MA: MIT Press.

Patel, A. D., and J. R. Iversen. 2014. The evolutionary neuroscience of musical beat perception: The action simulation for auditory prediction (ASAP) hypothesis. *Frontiers in Systems Neuroscience* 8:57. doi:10.3389/fnsys.2014.00057.

Patel, A. D., J. R. Iversen, M. R. Bregman, and I. Schulz. 2009. Experimental evidence for synchronization to a musical beat in a nonhuman animal. *Current Biology* 19 (10): 827–830. doi:10.1016/j.cub.2009.03.038.

Peretz, I. 2016. Neurobiology of congenital amusia. *Trends in Cognitive Sciences* 20 (11): 857–867. doi:10.1016/j.tics.2016.09.002.

Petkov, C. I., and E. D. Jarvis. 2012. Birds, primates, and spoken language origins: Behavioral phenotypes and neurobiological substrates. *Frontiers in Evolutionary Neuroscience* 4 (August): 12. doi:10.3389/fnevo.2012.00012.

Phillips-Silver, J., P. Toiviainen, N. Gosselin, O. Piché, S. Nozaradan, C. Palmer, and I. Peretz. 2011. Born to dance but beat deaf: A new form of congenital amusia. *Neuropsychologia* 49 (5): 961–969. doi:10.1016/j.neuropsychologia.2011.02.002.

Pinker, S. 1997. *How the Mind Works*. New York: Norton.

Porter, D., and A. Neuringer. 1984. Music discriminations by pigeons. *Journal of Experimental Psychology: Animal Behavior Processes* 10 (2): 138–148.

Ravignani, A., and P. F. Cook. 2016. The evolutionary biology of dance without frills. *Current Biology* 26 (19): R878–R879. doi:10.1016/j.cub.2016.07.076.

Reichmuth Kastak, C., and R. J. Schusterman. 2002. Long-term memory for concepts in a California sea lion (*Zalophus californianus*). *Animal Cognition* 5 (4): 225–232. doi:10.1007/s10071-002-0153-8.

Repp, B. H. 1991. Some cognitive and perceptual aspects of speech and music. In *Music, Language, Speech and Brain*, ed. J. Sundberg, L. Nord, and R. Carson, 257–268. Cambridge: Macmillan.

Richner, H. 2016. Interval singing links to phenotypic quality in a songbird. *Proceedings of the National Academy of Sciences of the United States of America* 113 (45): 12763–12767. doi:10.1073/pnas.1610062113.

Rizzolatti, G., and L. Craighero. 2004. The mirror-neuron system. *Annual Review of Neuroscience* 27:169–192. doi:10.1146/annurev.neuro.27.070203.144230.

Robbins, L., and S. W. Margulis. 2014. The effects of auditory enrichment on gorillas. *Zoo Biology* 33 (3): 197–203. doi:10.1002/zoo.21127.

Savage, P. E., S. Brown, E. Sakai, and T. E. Currie. 2015. Statistical universals reveal the structures and functions of human music. *Proceedings of the National Academy of Sciences of the United States of America* 112 (29): 8987–8992. doi:10.1073/pnas.1414495112.

Schachner, A., T. F. Brady, I. M. Pepperberg, and M. D. Hauser. 2009. Spontaneous motor entrainment to music in multiple vocal mimicking species. *Current Biology* 19 (10): 831–836. doi:10.1016/j.cub.2009.03.061.

Schusterman, R. J. 2008. Vocal learning in mammals with special emphasis on pinnipeds. In *The Evolution of Communicative Flexibility: Complexity, Creativity, and Adaptability in Human and Animal Communication*, ed. D. K. Oller and U. Gribel, 41–70. Cambridge, MA: MIT Press.

Schusterman, R. J., and R. F. Balliet. 1970. Conditioned vocalizations as a technique for determining visual acuity thresholds in sea lions. *Science* 169 (944): 498–501. doi:10.1126/science.169.3944.498.

Shannon, R. V. 2016. Is birdsong more like speech or music? *Trends in Cognitive Sciences* 20 (4): 245–247. doi:10.1016/j.tics.2016.02.004.

Shinozuka, K., H. Ono, and S. Watanabe. 2013. Reinforcing and discriminative stimulus properties of music in goldfish. *Behavioural Processes* 99:26–33. doi:10.1016/j.beproc.2013.06.009.

Sivic, J., B. Russell, and A. Efros. 2005. *Discovering Object Categories in Image Collections.* Computer Science and Artificial Intelligence Laboratory Technical Report. Cambridge, MA: Massachusetts Institute of Technology.

Spierings, M. J., and C. ten Cate. 2014. Zebra finches are sensitive to prosodic features of human speech. *Proceedings of the Royal Society B: Biological Sciences* 281 (1787): 1–7. doi:10.1098/rspb.2014.0480.

Tarr, B., J. Launay, and R. I. M. Dunbar. 2014. Music and social bonding: "Self–other" merging and neurohormonal mechanisms. *Frontiers in Psychology* 5:1–10. doi:10.3389/fpsyg.2014.01096.

ten Cate, C. 2017. Assessing the uniqueness of language: Animal grammatical abilities take center stage. *Psychonomic Bulletin and Review* 24:91–96. doi:10.3758/s13423-016-1091-9.

ten Cate, C., W. S. Bruins, J. den Ouden, T. Egberts, H. Neevel, M. J. Spierings, K. van der Burg, and A. W. Brokerhof. 2009. Tinbergen revisited: A replication and extension of experiments on the beak colour preferences of herring gull chicks. *Animal Behaviour* 77 (4): 795–802.

Tinbergen, N. 1951. *Study of Instinct.* Oxford: Clarendon Press.

Tinbergen, N. 1953. *The Herring Gull's World.* London: Collins.

Tinbergen, N., and A. Perdeck. 1950. On the stimulus situation releasing the begging response in the newly hatched herring gull chick (*Larus argentatus argentatus* Pont). *Behaviour* 3 (1): 1–39.

Tokarev, K., A. Tiunova, C. Scharff, and K. Anokhin. 2011. Food for song: Expression of c-Fos and ZENK in the zebra finch song nuclei during food aversion learning. *PLOS One* 6 (6): e21157. doi:10.1371/journal.pone.0021157.

Trehub, S. E. 2003. The developmental origins of musicality. *Nature Neuroscience* 6:669–673.

Ueno, A., S. Hirata, K. Fuwa, K. Sugama, K. Kusunoki, G. Matsuda, H. Fukushima, K. Hiraki, M. Tomonaga, and T. Hasegawa. 2008. Auditory ERPs to stimulus deviance in an awake chimpanzee (*Pan troglodytes*): Towards hominid cognitive neurosciences. *PLOS One* 3 (1): 5.

Vallet, E., and M. Kreutzer. 1995. Female canaries are sexually responsive to special song phrases. *Animal Behaviour* 49 (6): 1603–1610. doi:10.1016/0003-3472(95)90082-9.

van der Aa, J., H. Honing, and C. ten Cate. 2015. The perception of regularity in an isochronous stimulus in zebra finches (*Taeniopygia guttata*) and humans. *Behavioural Processes* 115:37–45. doi:10.1016/j.beproc.2015.02.018.

Velliste, M., S. Perel, M. C. Spalding, A. S. Whitford, and A. B. Schwartz. 2008. Cortical control of a robotic arm for self-feeding. *Nature* 453:1098–1101. doi:10.1038/nature06996.

Wang, A. L.-C. 2006. The Shazam music recognition service. *Communications of the ACM* 49:44–48. doi:10.1145/1145287.1145312.

Weisman, R. G., D. Mewhort, M. Hoeschele, and C. B. Sturdy. 2012. New perspectives on absolute pitch in birds and mammals. In *The Oxford Handbook of Comparative Cognition*, 2nd ed., ed. E. A. Wasserman and T. R. Zentall, 67–79. Oxford: Oxford University Press.

Weisman, R., M. Hoeschele, and C. B. Sturdy. 2014. A comparative analysis of auditory perception in humans and songbirds: A modular approach. *Behavioural Processes* 104:35–43. doi:10.1016/j.beproc.2014.02.006.

Whitham, J. C., M. S. Gerald, and D. Maestripieri. 2007. Intended receivers and functional significance of grunt and girney vocalizations in free-ranging female rhesus macaques. *Ethology* 113 (9): 862–874. doi:10.1111/j.1439-0310.2007.01381.x.

Winkler, I., G. P. Háden, O. Ladinig, I. Sziller, and H. Honing. 2009. Newborn infants detect the beat in music. *Proceedings of the National Academy of Sciences of the United States of America* 106 (7): 2468–2471. doi:10.1073/pnas.0809035106.

Zarco, W., H. Merchant, L. Prado, and J. C. Mendez. 2009. Subsecond timing in primates: Comparison of interval production between human subjects and rhesus monkeys. *Journal of Neurophysiology* 102 (6): 3191–3202. doi:10.1152/jn.00066.2009.

Zatorre, R. J., and V. N. Salimpoor. 2013. From perception to pleasure: Music and its neural substrates. *Proceedings of the National Academy of Sciences of the United States of America* 110 (supp.): 10430–10437. doi:10.1073/pnas.1301228110.

Index